Global Navigation Satellite Systems

Report of a Joint Workshop of the National Academy of Engineering and the Chinese Academy of Engineering

Edited by Lance A. Davis, Per K. Enge, and Grace X. Gao

NATIONAL ACADEMY OF ENGINEERING
OF THE NATIONAL ACADEMIES

THE NATIONAL ACADEMIES PRESS
Washington, D.C.
www.nap.edu

THE NATIONAL ACADEMIES PRESS 500 Fifth Street, NW Washington, DC 20001

This project was supported by funding from the National Academies and the National Academy of Engineering Fund. Any opinions, finding, or conclusions expressed in this publication are those of the workshop participants.

International Standard Book Number 13: 978-0-309-22275-4
International Standard Book Number 10: 0-309-22275-3

Copies of this report are available from the National Academies Press, 500 Fifth Street, NW, Keck 360, Washington, DC 20001; (888) 624-8373 or (202) 334-3313; *www.nap.edu*.

For more information about the National Academy of Engineering, visit the NAE home page at *www.nae.edu*.

Printed in the United States of America

THE NATIONAL ACADEMIES

Advisers to the Nation on Science, Engineering, and Medicine

The **National Academy of Sciences** is a private, nonprofit, self-perpetuating society of distinguished scholars engaged in scientific and engineering research, dedicated to the furtherance of science and technology and to their use for the general welfare. Upon the authority of the charter granted to it by the Congress in 1863, the Academy has a mandate that requires it to advise the federal government on scientific and technical matters. Dr. Ralph J. Cicerone is president of the National Academy of Sciences.

The **National Academy of Engineering** was established in 1964, under the charter of the National Academy of Sciences, as a parallel organization of outstanding engineers. It is autonomous in its administration and in the selection of its members, sharing with the National Academy of Sciences the responsibility for advising the federal government. The National Academy of Engineering also sponsors engineering programs aimed at meeting national needs, encourages education and research, and recognizes the superior achievements of engineers. Dr. Charles M. Vest is president of the National Academy of Engineering.

The **Institute of Medicine** was established in 1970 by the National Academy of Sciences to secure the services of eminent members of appropriate professions in the examination of policy matters pertaining to the health of the public. The Institute acts under the responsibility given to the National Academy of Sciences by its congressional charter to be an adviser to the federal government and, upon its own initiative, to identify issues of medical care, research, and education. Dr. Harvey V. Fineberg is president of the Institute of Medicine.

The **National Research Council** was organized by the National Academy of Sciences in 1916 to associate the broad community of science and technology with the Academy's purposes of furthering knowledge and advising the federal government. Functioning in accordance with general policies determined by the Academy, the Council has become the principal operating agency of both the National Academy of Sciences and the National Academy of Engineering in providing services to the government, the public, and the scientific and engineering communities. The Council is administered jointly by both Academies and the Institute of Medicine. Dr. Ralph J. Cicerone and Dr. Charles M. Vest are chair and vice chair, respectively, of the National Research Council.

www.national-academies.org

WORKSHOP STEERING COMMITTEE

BRADFORD W. PARKINSON, Edward C. Wells Professor of Aeronautics and Astronautics Emeritus, Stanford University
PER K. ENGE, Professor, Department of Aeronautics and Astronautics, Stanford University
LIU JINGNAN, Member, Chinese Academy of Engineering

Staff

LANCE A. DAVIS, Executive Officer, National Academy of Engineering
GRACE X. GAO, Rapporteur and Engineering Research Associate, Global Positioning System Laboratory, Stanford University
PENELOPE GIBBS, Senior Program Associate, NAE Program Office
PROCTOR P. REID, Director, NAE Program Office

U.S. DELEGATION

National Academy of Engineering

CHARLES M. VEST, President, National Academy of Engineering
LANCE A. DAVIS, Executive Officer, National Academy of Engineering
PER K. ENGE, Professor, Department of Aeronautics and Astronautics, Stanford University
BRADFORD W. PARKINSON, Edward C. Wells Professor of Aeronautics and Astronautics Emeritus, Stanford University
PROCTOR P. REID, Director, Program Office, National Academy of Engineering
CHARLES R. TRIMBLE, Chairman, U.S. Global Positioning System Industry Council

GPS Systems and Application Experts

PENINA AXELRAD, Professor, Department of Aerospace Engineering Sciences, Colorado Center for Astrodynamics Research, University of Colorado, Boulder
GRACE GAO, Engineering Research Associate, Global Positioning System Laboratory, Stanford University
RITA LOLLOCK, General Manager, Navigation Division, The Aerospace Corporation
MICHAEL O'CONNOR, Consultant and Principal, O'C and Associates
THOMAS D. POWELL, Systems Director, Navigation Division, The Aerospace Corporation
STUART RILEY, Manager, Signal Processing and Electronic Hardware Group, Engineering and Construction Division, Trimble Navigation
A.J. VAN DIERENDONCK, GNSS Consultant, AJ Systems
TODD WALTER, Senior Research Engineer, Department of Aeronautics and Astronautics, Stanford University

U.S. Government

LEO ELDREDGE, Manager, GNSS Group, Air Traffic Organization, Federal Aviation Administration
DAVID A. TURNER, Deputy Director, Space and Advanced Technology, U.S. Department of State

CHINESE DELEGATION

Chinese Academy of Engineering

ZHOU JI, President
BAI YULIANG, Secretary General
LIU JINGNAN, CAE member
SHEN RONGJUN, CAE member
WANG LIHENG, CAE member

Department of International Cooperation of CAE

XU JIN, Deputy Director-General
ZHENG XIAOGUANG, Assistant Director-General
WANG XIAOWEN, Deputy Director
YUAN MIN, Program Officer

Chinese Academy of Sciences

SUN JIADONG, CAS Member
YANG YUANXI, CAS Member

Research Institutes

CHEN JINGPING, Senior Engineer, Beijing Global Information Center of Application and Exploitation
DING QUN, Researcher, Xi'an Research Institute of Navigation
DING XIANCHENG, Researcher, China Electronics Technology Group Corporation
DU XIAODONG, Researcher, Beijing Research Institute of Telemetry
LI ZUHONG, Researcher, China Academy of Space Technology
LU XIAOCHUN, Researcher, National Time Service Center, Chinese Academy of Sciences
TAN SHUSEN, Researcher, Beijing Global Information Center of Application and Exploitation
WANG LI, Researcher, China Aerospace Science and Technology Consultant Corporation
WU HAITAO, Researcher, Satellite Navigation Headquarters, Chinese Academy of Sciences

Corporation

HAN SHAOWEI, Researcher, Unicore Communications Inc.

Chinese Government

CAI LANBO, Vice Director, China Satellite Navigation Office
GUO SHUREN, Researcher, China Satellite Navigation Office
LU JUN, Senior Engineer, China Satellite Navigation Office
RAN CHENGQI, Director, China Satellite Navigation Office
YANG CHANGFENG, Deputy Chief Designer of Beidou Navigation Satellite System
YANG JUN, Vice Director, China Satellite Navigation Office

Industry Administrative Departments

BAI SHULIN, Deputy Director, Joint Center of GNSS, Ministry of Education, China, and Deputy Director, Advanced Technology Institute, Peking University
LU XIAOPING, Inspector, Civil Aviation Administration of China

Preface

The Global Positioning System (GPS) has revolutionized the measurement of position, velocity, and time. It has rapidly evolved into a worldwide utility with more than a billion receiver sets currently in use. Applications include emergency services, aircraft landing, farm tractor auto steering, and measurements of Earth tectonic motion, with new applications appearing every year. There are enormous benefits to humanity: improved safety of life, increased productivity, and widespread convenience. Recognizing this utility, Russia is rejuvenating its satellite navigation system, GLONASS. Other nations are now deploying new Global Navigation Satellite Systems (GNSSs) of their own: the European Galileo system and the Chinese System called Compass or BeiDou. Regional satellite systems are also being deployed: Japan's Quasi-Zenith Satellite System (QZSS) and the Indian Regional Navigation Satellite System (IRNSS). All of humanity will benefit if these systems can operate as one super system, with users able to navigate using any four satellites.

The GPS constellation may decline in size over the next five years (from the current 31 satellites). This will reduce signal availability for those GPS-only users whose views of the sky are restricted by mountains, canopy, and/or city buildings. An expanded constellation of satellites providing more signals would significantly improve system access and reliability for these sky-impaired users. Moreover, a frequency diversity of signals provided by an expanded constellation would help mitigate inadvertent radio frequency interference with the weak GPS signal.

The deployment of the new and modernized satellite systems has the potential to greatly reduce signal outages—but only if their system and signal specifications are well understood, interoperable, and officially supported so that receivers can be manufactured and their integrity assessed. The Russians and Europeans have provided significant information to the United States with respect to their signal and systems details. At the time of the workshop, held on May 24 and 25, 2011, in Shanghai, China, the Chinese authorities had not released an official description of its civil signals, and many of the U.S. delegates expressed a wish for this critical information. As the report was going to press, China released this crucial document in December 2011, seven months after the workshop described herein occurred.

Our bilateral workshop's goal was to promote technical and policy-related cooperation between the United States and China regarding their respective navigation satellite systems—the U.S. Global Positioning System and the Chinese Compass system—to the benefit of China, the United States, and GNSS users worldwide. The workshop sought to encourage greater transparency regarding the technical and operational details that would allow the two countries to exchange system- and signal-level specifications. The recent exchange of data will improve the accuracy and availability of real-time position, navigation, and time data for all users worldwide. This exchange will foster the interchangeability of satellite signals, which will greatly decrease outages, particularly for sky-impaired users.

The workshop organizers would like to thank the National Academy of Engineering and the Chinese Academy of Engineering for their support of this workshop. We would also like to thank all the workshop participants for their thoughtful presentations and discussion.

Bradford W. Parkinson Per K. Enge Liu Jingnan

Acknowledgments

This summary has been reviewed in draft form by individuals chosen for their diverse perspectives and technical expertise, in accordance with procedures approved by the National Academies. The purpose of the independent review is to provide candid and critical comments to assist the NAE in making its published report as sound as possible and to ensure that the report meets institutional standards for objectivity, evidence, and responsiveness to the study charge. The review comments and draft manuscript remain confidential to protect the integrity of the deliberative process. We wish to thank the following individuals for their review of this report:

Charles Trimble, GPS Industry Council
Charles Elachi, Jet Propulsion Laboratory
Thomas Herring, Massachusetts Institute of Technology
Anthony Russo, National Coordination Office for Space-Based Positioning, Navigation and Timing

Although the reviewers listed above have provided many constructive comments and suggestions, they were not asked to endorse the views expressed in the report, nor did they see the final draft of the report before its release. The review of this report was overseen by Julia M. Phillips, Sandia National Laboratories. Appointed by NAE, she was responsible for making certain that an independent examination of this report was carried out in accordance with institutional procedures and that all review comments were carefully considered. Responsibility for the final content of this report rests entirely with the authors and NAE.

Contents

APPENDIXES

Summary of the Workshop

As Reported by
GRACE XINGXIN GAO

This report summarizes the joint workshop on Global Navigation Satellite Systems (GNSS) held jointly by the U.S. National Academy of Engineering (NAE) and the Chinese Academy of Engineering (CAE) on May 24–25, 2011, at Hongqiao Guest Hotel in Shanghai, China.

The workshop had three objectives. First, it sought to explore issues of enhanced interoperability and interchangeability for all civil users. Second, it aimed to consider collaborative efforts for countering the global threat of inadvertent or illegal interference to GNSS signals. Third, it intended to promote new applications for GNSS, emphasizing productivity, safety, and environmental protection.

The workshop presentations were chosen based on the following criteria: the presentations must have relevant engineering/technical content or usefulness; be of mutual interest; offer the opportunity for enhancing GNSS availability, accuracy, integrity, and/or continuity; and offer the possibility of recommendations for further actions and discussions.

Dr. Wang Liheng, member of CAE, started the opening ceremony of this workshop. Dr. Zhou Ji, president of CAE, and Dr. Charles M. Vest, president of NAE, provided opening remarks. The government delegates from both sides introduced the evolution process, current status, future development, applications, and international cooperation of the BeiDou/Compass Navigation Satellite system and the Global Positioning System (GPS). Experts and scholars from both the U.S. and Chinese delegations exchanged knowledge and discussed GNSS compatibility and interoperability, system monitoring and service improvement, satellite navigation terminal and application technology, as well as talent exchanges.

The key outcomes of the workshop follow.

SHARED HISTORY AND STATUS OF GPS AND COMPASS

Mr. David Turner, deputy director of space and advanced technology, U.S. Department of State, provided an update on GPS constellation status and current U.S. government GPS policy, programs, and international activities. His comments are now summarized.

At the time of the workshop, the GPS constellation consisted of 31 operational satellites, although the baseline is a constellation of 24 satellites. These operational satellites included: 11 GPS Block IIA satellites, 12 GPS Block IIR satellites, 7 GPS Block IIR-M satellites, and 1 GPS Block IIF satellite. The GPS IIA satellites initiated the Block II series. Nineteen GPS IIA satellites were launched—the first on November 26, 1990, and the last on November 6, 1997. The Block IIR series are "replenishment" satellites developed by Lockheed Martin. Each satellite weighs 4,480 pounds (2,030 kg) at launch and 2,370 pounds (1,080 kg) once on orbit. The first successful IIR launch was in July 1997. The Block IIR-M satellites are modernized IIR satellites that include a new military signal and a second civil signal, known as L2C. The first IIR-M satellite was launched in September 2005. The Block IIF series are "follow-on" satellites developed by Boeing. Boeing is under contract to build a total of 12 Block IIF satellites. The Block IIF provides a third frequency for civil use (L5), allowing position determinations with even higher precision. The first Block IIF satellite was launched in May 2010.

Looking forward, GPS modernization will result in a variety of new signals, including civil signals L2C, L5, and L1C, all at different frequencies. L2C contains two distinct pseudo random noise (PRN) code sequences to provide ranging information—the Civilian Moderate length code (CM) and the Civilian Long length code (CL). The CM code is 10,230 bits long, repeating every 20 ms. The CL code is 767,250 bits long, repeating every 1,500 ms. Each signal is transmitted at 511,500 bits per second (bits/s); however, they are multiplexed together to form a 1,023,000 bits/s signal. CM is modulated with the navigation messages, whereas CL does not contain any modulated data and is called a dataless sequence. The L5 signal also contains two PRN ranging codes: the in-phase code (denoted as the I5-code); and the quadrature-phase code (denoted as the Q5-code). Both codes are 10,230 bits long and transmitted at 10.23 MHz.

The L1C civil signal is a more robust signal broadcast on the legacy L1 frequency (1575.42 MHz), which contains the coarse/acquisition (C/A) signal used by all current GPS users. The PRN codes are 10,230 bits long and transmit at 1.023 Mbps. It uses both pilot and data carriers like L2C. The modulation techniques used are BOC(1,1) for the data signal and TMBOC (time multiplexed binary offset carrier) for the pilot. Of the total L1C signal power, 25 percent is allocated to the data and 75 percent to the pilot.

GPS performance is better than ever and will continue to improve. U.S policy encourages worldwide use of civil GPS and augmentation systems. International cooperation is a priority in pursuit of systems compatibility and interoperability with GPS.

With regard to the Compass system, both Drs. Ran Chengqi, director, China Satellite Navigation Office, and Tan Shusen, a researcher with the Beijing Global Information Center of Application and Exploitation, stated that the Compass Navigation system will follow a "three-step" deployment, from regional to global, and from active to passive. The Compass system aims to achieve regional passive service coverage by 2012 and global passive service coverage by 2020.

The first step occurred in the 1980s when China began to conduct a study on the BeiDou demonstration system under a dual-satellite active positioning mechanism. First initiated in 1994, the BeiDou demonstration system was able to provide regional active services in 2000. From 2000 to 2003, three satellites were launched. In 2010, these early satellites were replaced with newer ones.

The second step started in 1999. China began to study the BeiDou navigation satellite system under a passive positioning mechanism. System construction was formally started in 2004. The system will provide Regional Navigation Satellite Service (RNSS) for users in China and its surrounding areas by 2012.

For the third step, the BeiDou system will finally provide compatible and interoperable global coverage and also upgrade its regional service by 2020.

INTEROPERABILITY AND INTERCHANGEABILITY OF GPS AND COMPASS

Compatible constellations do not interfere with each other, and interoperable constellations gain strength from each other. Thus interoperability is a stronger condition than compatibility. If the GNSS signals are interoperable, then a receiver that combines the signals enjoys improved availability relative to a receiver that employs only one constellation. With this property, a receiver in an obstructed signal environment (urban, forest, or mountains) is more likely to have the number of satellites needed for good performance. Accuracy also improves and integrity improvements based on multiple constellations are under active investigation.

Interchangeability is a still stronger and more specific requirement. It enables the user to estimate the four estimanda (latitude, longitude, altitude, and user time offset) based on any four satellites from any of the cooperating constellations. Interchangeability requires that all constellations have a synchronized clock and operate on a common (or nearly common) carrier frequency.

To encourage interoperability and interchangeability, the service providers must enable a diverse group of receiver manufacturers to innovate and build equipment that takes advantage of the multiplicity of offered constellations. Specifically, they must provide the interface control documents (ICDs) for the civil signals to the receiver manufacturers well in advance of the constellation being operational. The GPS program has been well served by this openness, and the Chinese delegates mentioned that the ICD for Compass may be available as soon as December of 2011. And, as this report was going to press, China did release the ICD for its civil signals in December of 2011.

In his talk, Dr. Thomas Powell, systems director of the navigation division for The Aerospace Corporation, discussed compatibility and interoperability. He described the benefits to GNSS providers, receiver manufacturers, and users. As a result of international bilateral and multilateral meetings, the advantages of GNSS interoperability have been recognized and major strides have been made toward achieving the goal. Specifically, the international community has agreed on right-hand circular polarization and code division multiple access (CDMA) modulation. In addition, GPS, Compass, Galileo, and the Japanese QZSS system use two identical GNSS carrier frequencies. GLONASS is also considering this possibility.

To achieve interoperability, multiple GNSS systems should have compatible or consistent system time and geodetic coordinate systems. The Compass/BeiDou satellite navigation system follows the rules of compatibility and interoperability defined by the International Committee of GNSS (ICG). The coordinate system of BeiDou is aligned to the China Geodetic Coordinate System 2000, which is aligned to the International Terrestrial Reference System (ITRS). BeiDou system time (BDT) is an internal, continuous navigation time scale, without leap second. BDT is linked to the national Coordinated Universal Time (UTC).

In the upcoming decade, GPS signals will be modernized, the Russian GLONASS satellite constellation will be rejuvenated, and the following new systems will become operational: Chinese Compass, European Galileo, and Japanese QZSS. These modernized or new systems will provide three GNSS frequencies. For precision applications, they will enable 10 cm accuracy over wide areas. Availability and accuracy for all forms of navigation will improve, because the number of visible satellites will more than double, assuring users will have access to GPS, Compass, and other GNSS systems.

Professor Per Enge of Stanford University provided a briefing on "Global Safety of Life Service from Multiple GNSS Constellations" (e.g., GPS + Compass). By comparing today's single-frequency technology for detection of GPS faults and the future utility of dual frequencies and multiple constellations, he showed that multi-constellation GNSS will bring benefits to aviation, especially in terms of integrity.

MUTUAL AVAILABILITY OF SIGNAL SPECIFICATIONS

Mr. Turner described the U.S. policy promoting global use of GPS technology. There are no direct user fees for civil GPS services. The GPS service is provided on a continuous and worldwide basis. Moreover, the GPS signal definitions and structures are open and public for all civil services.

Dr. Ran stated that the Chinese Compass system will also ensure the openness and availability of its civil signals. In general, China aims to promote equal access for user equipment manufacturing, applications development, and value-added services.

ESTABLISHING A GLOBAL PERFORMANCE MONITORING AND ASSESSING SYSTEM

GNSS would benefit from worldwide monitoring and assessment of the civil signals (i.e., open services). Such monitoring and visibility would extend the depth and breadth of GNSS applications and also provide the foundation for civil users to utilize system services safely and reliably. Therefore, monitoring of the GNSS Open Services has become a focus of attention for all providers and users of GNSS. Such open assessment is particularly valuable during the construction and development process of a particular GNSS constellation. Different GNSS providers and resources from different regions worldwide are involved in this issue, and enhancing international exchanges and cooperation is vital.

Mr. Ding Qun, a researcher with Xi'an Research Institute of Navigation, proposed an international GNSS Monitoring and Assessment System (iGMAS) for monitoring the performance of multiple GNSS systems. iGMAS is divided into the following layers: constellation state layer, spatial signals layer, navigation information layer, and service performance layer. The basic functions of iGMAS include data monitoring and collection, data transmission, data storage, data analysis, and information release.

INTEROPERABILITY OF THE U.S. AND CHINESE SPACE-BASED AUGMENTATION SYSTEMS (SBAS)

GNSS space-based and ground-based augmentation systems effectively improve the performance of satellite navigation and positioning services. Such systems are commonly composed of multiple ground stations, located at accurately surveyed points. The ground stations take multiple measurements of the GNSS satellites in view. Using these measurements and its known location, the ground system creates corrections and real-time error bounding information for broadcast to the end users. The corrections improve accuracy, continuity, and availability. The real-time error bounds protect the integrity of the navigation system. Compass and GPS would benefit from enhanced exchanges and cooperation regarding space-based and ground-based augmentation systems.

According to the Chinese delegation, the Compass system consists not only of medium Earth orbit (MEO) satellites, but also at least three Geostationary (GEO) satellites. The Chinese Compass GEO satellites will serve both as the satellites for ranging and for space-based augmentation. In contrast, the United States does not presently place any GPS satellites in geostationary orbit, nor does the United States broadcast aviation safety information from GPS satellites. Rather, it places transponders on commercial geostationary satellites for space-based augmentation.

In addition, the Chinese space-based augmentation system may provide cross-constellation integrity assistance. That is, the Chinese GEO satellites may broadcast error corrections messages for the Chinese Compass system and GPS and other GNSS systems. The United States has no immediate plans to broadcast error corrections for other constellations.

Certification of GNSS for use in aircraft navigation has already taken place for GPS L1 signals in many parts of the world. Even so, certification remains a challenging enterprise. The associated safety analyses rely on prototyping and careful analysis as starting points. These data and studies are used to generate and validate the threat models needed to support certification. New regions seeking to certify GNSS would benefit by drawing upon the lessons learned from previous experiences.

Dr. Todd Walter, senior research engineer at Stanford University, shared the lessons learned from developing the U.S. Wide Area Augmentation System (WAAS). Chief among these was the use of threat models. Threat models define the fault modes, how they manifest themselves, and how likely they are to occur. They describe the feared events that one must protect against. A well-defined threat model permits a quantitative assessment of the mitigation strategy. A quantitative assessment as opposed to a qualitative assessment is essential to establishing integrity. Certifying that WAAS is safe for instrument flight rules (IFR) (i.e., flying in bad weather) requires proving that there is only an extremely small probability that an error exceeding the requirements for accuracy will go undetected. Specifically, the probability is stated as 1×10^{-7} (or 10^{-7} integrity) and is equivalent to no more than 3 seconds of undetected faulty data per year.

Another key lesson is the application of the 10^{-7} integrity requirement to each approach. Rather than averaging over conditions with different risk levels, one must overbound the conditions describing the worst allowable time and place.

Dr. Grace Xingxin Gao, engineering research associate at Stanford University, addressed SBAS collaboration from the perspective of navigating in the Arctic and analyzed techniques to extend SBAS coverage to this critical area. She investigated three modifications to the SBAS architectures used today. First, an Arctic SBAS could use data from the existing ground networks for the SBAS in the United States, Europe, and Russia. These reference networks could be augmented by new reference stations located in the Arctic. Second, an Arctic SBAS would not be able to use geostationary satellites to broadcast the data to Arctic users, but could use low Earth orbiting satellites (LEOs, e.g., Iridium) to broadcast the corrections and integrity data to the end users. Third, multiple GNSS constellations (e.g., GPS + Compass) would significantly reduce VDOPs (vertical dilution of precision) and thus would improve vertical accuracy in the Arctic. Subject to the resolution of technical issues (e.g., high Doppler rates, synchronization of the time of transmission), the LEO satellites could also be used as ranging sources.

SUPPRESSION OF WORLDWIDE CIVIL SIGNAL JAMMING DEVICES

The Compass and GPS satellites are in medium Earth orbit, and so the received signals are very weak. These signals are easily overcome by radio frequency interference (RFI) from terrestrial sources. During RFI events, the safety and efficiency benefits of GNSS are lost. Thus, the worldwide community must cooperatively pursue an interference-free environment for the GNSS signals. Technical discussions were conducted regarding interference identification and localization. Jamming resistant solutions were explored from system design, terminal applications, and other broader aspects.

Civil signals face two kinds of interference: intentional and unintentional. Unintentional interference sources can be further categorized:

- Radio systems in or near the GNSS frequency band.
- Harmonics or intermodulation products produced by radio broadcasting and communication emitters.
- Stray signals from electronic equipment into the navigation frequency band.

Of particular concern, the delegates discussed the January 2011 conditional authorization by the U.S. Federal Communication Commission for LightSquared (LSQ) Corporation to deploy 40,000 powerful (up to 15 kW) broadband transmitters in radio spectrum previously reserved for weak space-to-Earth signals adjacent to the faint GPS signal. With a transmitter spacing of 400 to 800 m in cities, the LSQ signal would be more than 5 billion times more powerful than the GPS signal. Hence, the proposed LSQ signals are dangerous to the current GNSS system. Moreover, the LSQ band is even closer to the Compass B1 band, and so it may cause even more serious interference to Compass.

As discussed by Dr. A.J. Van Dierendonck of AJ Systems, intentional jammers, especially personal privacy devices (PPDs), have become an important issue for GNSS. PPDs are small and cheap jammers that can plug into cigarette lighter sockets in a vehicle. They are designed to "jam" or "mask" GPS (GNSS) receiver/navigators that report, through other means, the position of the host vehicle when the user does not want his or her position known. Use of PPDs is illegal in the United States, although they are usually available for purchase via the Internet. Unknown (or known) to the owner, PPDs can jam other users of GPS (GNSS), including Safety-of-Life (SOL) operations. At Newark Liberty International Airport (EWR) in New Jersey, the RFI from such a device affected the operation of the Ground-Based Augmentation System (GBAS). GBAS performance interruptions have been observed since November 2009, impacting GBAS service availability. The PPDs do not seem to directly interfere with the airborne avionics, but as they interrupt the data from the GBAS ground system, they could interrupt the continuity of the aircraft approach operation.

Mr. Leo Eldredge, manager of the GNSS Group of the U.S. Federal Aviation Administration (FAA), presented the need for robust radio navigation and potential architectures for alternative position, navigation, and timing (APNT). The FAA has concentrated on three categories of solutions that appear promising, while inviting input from the public and industry at meetings, symposia, and conferences on other potential areas of research. The three categories currently being considered are Optimized Distance Measuring Equipment (DME) Network, Wide-Area Multi-lateration, and a Pseudolite Network.

Mr. Du Xiaodong, a researcher at the Beijing Research Institute of Telemetry, stated that governments have the responsibility to protect the navigation frequency band against illegal interference. China has clear legal provisions for the usage and protection of radio spectrum resources and the investigation, production, distribution, and importation of radio equipment. Relevant laws and regulations are: *Real Right Law of the People's Republic of China*, *Criminal Law of the People's Republic of China*, *Radio Regulations of the People's Republic of China*, and *Radio Station License Regulations*. The protection of navigation signals spectrum resources relies on technology and the effective implementation of spectrum management. The China Radio Administration Bureau of Industry and Information Ministry has overall responsibility for administration and coordination. Its responsibilities are frequency spectrum monitoring, interference detection and investigation, matters of electromagnetic interference coordination, and maintenance of the transmission of radio waves in the air.

Receivers with the capability of tracking multiple GNSS systems may achieve better accuracy and integrity compared to those that track a single GNSS system. Dr. Han Shaowei from Unicore Communications Inc. showcased the Compass/GPS dual constellation receiver developed by his company. The receiver is a deeply integrated receiver, which has been used in China in precise positioning and also navigation applications. Dr. Han demonstrated the advantages of the Compass/GPS coupled receiver for sky-impaired users, such as those in an urban environment.

APPLICATIONS FOR GNSS

GNSS applications have made significant contributions to science, engineering, and commerce. They require the public availability of ICDs for the underlying GNSS systems. The success of GPS is based on industry-funded innovation combined with the open availability of these public documents. With ICDs, the forthcoming set of interoperable satellite constellations will improve the utility of current applications and open new sectors for exploration. Without ICDs, this expansion would have been limited.

Dr. Penina Axelrad, professor at the University of Colorado, Boulder, presented the application of GNSS to environmental studies. GNSS signals are influenced by the transmission media and interaction with surfaces near the receiving antenna. Observation of the modified signals from the ground and from airborne and space-borne platforms allows for scientific study of the ionosphere, atmosphere, and the Earth's surface. With appropriate signal processing, standard ground-based receivers provide estimates of atmospheric water vapor and soil moisture. Specialized receivers can measure occulted signals and enable high-resolution estimates of atmospheric density. Receivers measuring reflected signals are used to infer surface roughness and reflectivity, which can be related to surface conditions like ocean winds, soil moisture, and ice type. Modern receivers that can use multiple GNSS constellations will provide a rich global data set for environmental study.

Dr. Michael O'Connor of O'C and Associates briefed attendees on the recent advances of applying GNSS for precision agriculture. Precision agriculture (or precision farming) uses technology to better measure and control crop production on a site-specific basis to improve efficiency. Examples include more efficient application of seed and fertilizer and more effective utilization of tillage equipment. GNSS has been a key enabling technology for precision agriculture since the mid-1990s. Dr. O'Connor indicated that adoption of precision agriculture for seed and fertilizer management will improve when three key challenges have been overcome: (1) the improvement of GNSS signal availability, (2) the improvement of the efficiency of soil measurements, and (3) the wider adoption of simultaneous data analysis across multiple farms.

Dr. Liu Jingnan, CAE member, presented China's effort to use Precision Point Positioning (PPP) to monitor earthquakes. With high-rate GNSS data, both seismic waveforms and permanent offsets can be observed. Combining high-rate GNSS data and seismometer data, seismologic research on earthquake parameter determination and fault rupture modeling will be more scientific and reliable. Combined data can be applied for real-time earthquake monitoring, tsunami warning, and other engineering deformation monitoring.

Dr. Tan presented a variety of Compass applications. First, he discussed a dispatching system for the marine fishery. China has realized offshore real-time monitoring of about 2,000 fishing vessels by taking advantage of Compass for positioning, location reporting, and short message communication. Other applications include: a hydrological data collection system at the Three Gorges of the Yangtze river; a detection system for forest fires; an inspection and monitoring system for high-voltage power lines in remote areas; a system for environmental protection and energy conservation in shipping on the Yangtze River; a vehicle navigation system for all vehicles to share location information and reduce the severity of traffic jams; and a two-way timing service that provides a standard of high accuracy time and frequency for upgrading the old power grid.

TALENT EXCHANGE

The U.S. and Chinese delegations discussed the benefits that could accrue to both sides by promoting a deep and multilevel cooperation and exchange program regarding GNSS. Key observations from the workshop include the following: (1) it is beneficial for technology advancement and construction of the BeiDou system and GPS system to strengthen technical exchanges and cooperation between CAE and NAE in the GNSS field; (2) it is valuable to enhance academic exchanges and mutual understanding between technical personnel on both sides to improve technologies together; and (3) such academic exchange activities may be organized regularly or irregularly as needed.

Talent exchanges could include: consulting services; technology and scientific exchange, such as organizing technical conferences; and training from both sides. Specific activities may include: China inviting U.S. experts to give seminars to Chinese researchers and engineers; the United States inviting Chinese experts to visit U.S. institutions and enterprises and to communicate with U.S. experts; and the sending and receiving of visiting students and scholars.

SUMMARY

In conclusion, the U.S. and Chinese delegates experienced a very useful exchange of knowledge and experience in the field of GNSS. As Dr. Vest stated in his opening remarks, "We have one world, and only one set of global resources. It is important to work together on satellite navigation. Competing and cooperation is like Yin and Yang. They need to be balanced." This NAE-CAE Workshop on GNSS achieved the balance between such Yin and Yang of the U.S. and Chinese sides and has hopefully provided a basis for mutually beneficial and fruitful cooperation in the future.

The reader is invited to review papers from the conference presented in the following pages.

Address at the Opening Ceremony of the NAE-CAE Joint Workshop on Global Navigation Satellite Systems (GNSS)

ZHOU JI
President
Chinese Academy of Engineering

Dear President Vest,
Respected Members, experts,
Ladies and gentlemen,

On behalf of the Chinese Academy of Engineering, I would like to extend warm congratulations on the convening of the Workshop on Global Navigation Satellite Systems and a heartfelt welcome to all both from China and the United States.

As we all have witnessed, China has achieved great and historical achievements ever since it carried out the reform and opened up. At present, China's development is reaching a critical turning point. The Chinese Government clearly stated that China will go ahead under the theme of scientific development, and its main line will focus on accelerating the transformation of the mode of economic development. The acceleration of the transformation fundamentally counts on the force of science and technology, and the most critical task is to improve an independent innovation capability.

In the course of speeding up the transformation, China, on the one hand, will have to make good use of advanced engineering technology, remake and upgrade traditional industries, build up a modern industry system, and thus fundamentally enhance the overall scientific and technological quality and integrated competitiveness of the industries. On the other hand, China will have to cultivate and develop new industries of strategic importance and foster new sources of economic growth while taking innovation as a driving principle. In 2010, the Chinese Government enacted the *Decision of Speeding Up the Cultivation and*

Development of Strategic Emerging Industries. The Decision identifies seven major strategic emerging industries to be fostered and developed in the present stage: energy saving and environmental protection, new-generation information technology, biology, high-end equipment manufacturing, new energies, advanced materials, and new energy vehicles.

The Global Navigation Satellite System and its applications will constitute an important component of strategic emerging industries. As an important space infrastructure, GNSS will bring huge social and economic benefits to human beings. This technology is now increasingly penetrating into human life, society, and economy. Given the fact that China is a country with a large population and vast territory, to develop GNSS will make contributions to the enhancement of Chinese living standards and to the establishment of a harmonious society. It is of great significance to promote the development of China's strategic emerging industries.

Chinese culture values harmony as most precious and seeks harmony but not uniformity. For example, in Chinese families, we believe "harmony brings wealth" and, with regard to relationships, such as those between individuals or between nations, Chinese people value exchanges, cooperation, and a win-win. That is also the essence of the Chinese culture, which is deeply rooted in the consciousness of Chinese people and is reflected in the concept of scientific development. We expect a peaceful development, and we try to closely integrate China's development with the common prosperity of all other countries and jointly build a harmonious world.

In January 2011, Chinese President Hu Jintao paid a visit to the United States and carried the friendship from 1.3 billion Chinese to the American people. The visit fully demonstrated the Chinese people's good will to strengthen dialogues, enhance mutual trusts, expand exchanges, and deepen cooperation. The United States and China have signed cooperative agreements in a number of scientific and technological fields that involve energy-efficient buildings, clean coal, and electric vehicles. They will become new chapters in scientific and technological cooperation between the two countries.

Centering on the issues of engineering, science, and technology of mutual interests, the Chinese Academy of Engineering would like to further strengthen exchanges and cooperation with the U.S. National Academy of Engineering, continue to improve cooperative levels, expand cooperative fields and scopes, and contribute intelligence and strength to the friendship between the two countries and to the prosperity of the whole world and human civilization.

Finally, I wish the workshop a complete success!

Introductory Remarks

CHARLES M. VEST
President
National Academy of Engineering

On behalf of the U.S. National Academy of Engineering (NAE), I want to thank the Chinese Academy of Engineering (CAE) for hosting this joint workshop on the future of Global Navigation Satellite Systems (GNSS).

This is one of several joint activities our two academies have undertaken. We are especially proud of our work together to establish the ongoing China–U.S. Frontiers of Engineering symposia to advance knowledge and personal relationships across engineering disciplines and between nations. We also look forward to our continuing work together on such important topics as renewable energy and synthetic biology.

In today's world with its globalized economies and common challenges, great nations and organizations both compete with each other and cooperate with each other. Competition drives excellence, but increasingly cooperation is even more important, especially between organizations such as ours that are dedicated to advancing technology for the common good of humanity. We have one world, and only one set of global resources. It is important to work together on satellite navigation. Competing and cooperation is like Yin and Yang. They need to be balanced.

The NAE hopes that this workshop will stimulate greater cooperation between our two countries as we deploy and refine our GNSS systems and operations. Providing broader and more redundant coverage and operation would help GNSS provide better tools for science and technology, ranging from precise measurements of geological motions to more efficient agriculture that requires less water and minimizes fertilizer use. We also hope that we can stimulate cooperation in guarding against inappropriate interference of other signals with GNSS, and that further scientific and engineering interactions between our respective experts will follow this workshop. Above all, we hope that we can work toward maximum benefits for human safety that GNSS can provide.

The NAE very much appreciates the CAE's generous hosting of this workshop. Our delegation all looks forward to learning and working together with you.

WORKSHOP PRESENTATIONS

Development of the BeiDou Navigation Satellite System

RAN CHENGQI
China Satellite Navigation Office

INTRODUCTION OF THE BEIDOU NAVIGATION SATELLITE SYSTEM

Development Programme

The general development guideline of the BeiDou Navigation Satellite System (hereinafter in short the BeiDou system) includes quality, safety, application, and benefits. As a global satellite navigation system, the BeiDou system aims to provide all-day and all-weather navigation services of high performance and reliability for users around the world, with regional passive services coverage by 2012 and global passive services coverage by 2020.

The BeiDou system adopts four basic principles of openness, independence, compatibility, and gradualness.

Openness implies that the BeiDou system will provide high-quality open service free of charge for worldwide users and communicate with other countries to facilitate the development of GNSS technologies and industry.

Independence means that China will develop and operate the BeiDou system independently, provide services independently for worldwide users, and in particular provide high-quality services in the Asia-Pacific region.

Meanwhile, the BeiDou system will pursue solutions to realize compatibility and interoperability with other satellite navigation systems so that users can get better services with interoperable receivers, and so that the system can improve backward compatibility, ensuring the better interests and services of users and industries.

Based on the gradualness principle, the BeiDou system will follow a step-by-step pattern in accordance with technical and economic evolution in China, provide long-term continuous services for users, improve system performance, and ensure seamless and smooth transition between the system construction phases.

Figure 1 shows the overall development planning of the BeiDou system: "three-step" deployment, from regional to global, from active to passive.

First Step

In the 1980s, in line with the condition and demands of the time, China began to conduct a study on the BeiDou demonstration system under dual-satellite active positioning mechanism. First initiated in 1994, the BeiDou demonstration system was able to provide regional active services in 2000. From 2000 to 2003, three satellites were launched, and in 2010, these satellites have been replaced by new ones.

Second Step

In 1999, China began to study the BeiDou navigation satellite system under passive positioning mechanism. System construction formally started in 2004. The system will provide radio navigation satellite service (RNSS) for users in China and its surrounding areas on the basis of radio determination satellite service (RDSS) from 2012.

Third Step

The BeiDou system will finally provide compatible and interoperable RNSS with global coverage and also upgraded RDSS by 2020.

The BeiDou system is comprised of three major components: space segment, ground segment, and user segment. The space segment consists of 5 geostationary (GEO) satellites and 30 non-GEO satellites. The ground segment consists of several master control stations (MCS), upload stations (US), and a network of globally distributed monitor stations (MS). The user segment includes BeiDou terminals and interoperable terminals compatible with other GNSS systems, to meet different application requirements from different fields and industries.

The BeiDou system can transmit multiple navigation signals. Generally speaking, the system can provide two kinds of global services: open service and authorized service. Open service is free to users with positioning accuracy better than 10 m, timing accuracy better than 20 ns, and velocity accuracy of 0.2 m/s. The system can also provide two kinds of regional services—wide-area differential service and short message service. Wide-area differential accuracy is better than 1 m.

The time reference for the BeiDou system uses BeiDou time (BDT), which is traceable to the Coordinated Universal Time (UTC) maintained by the National

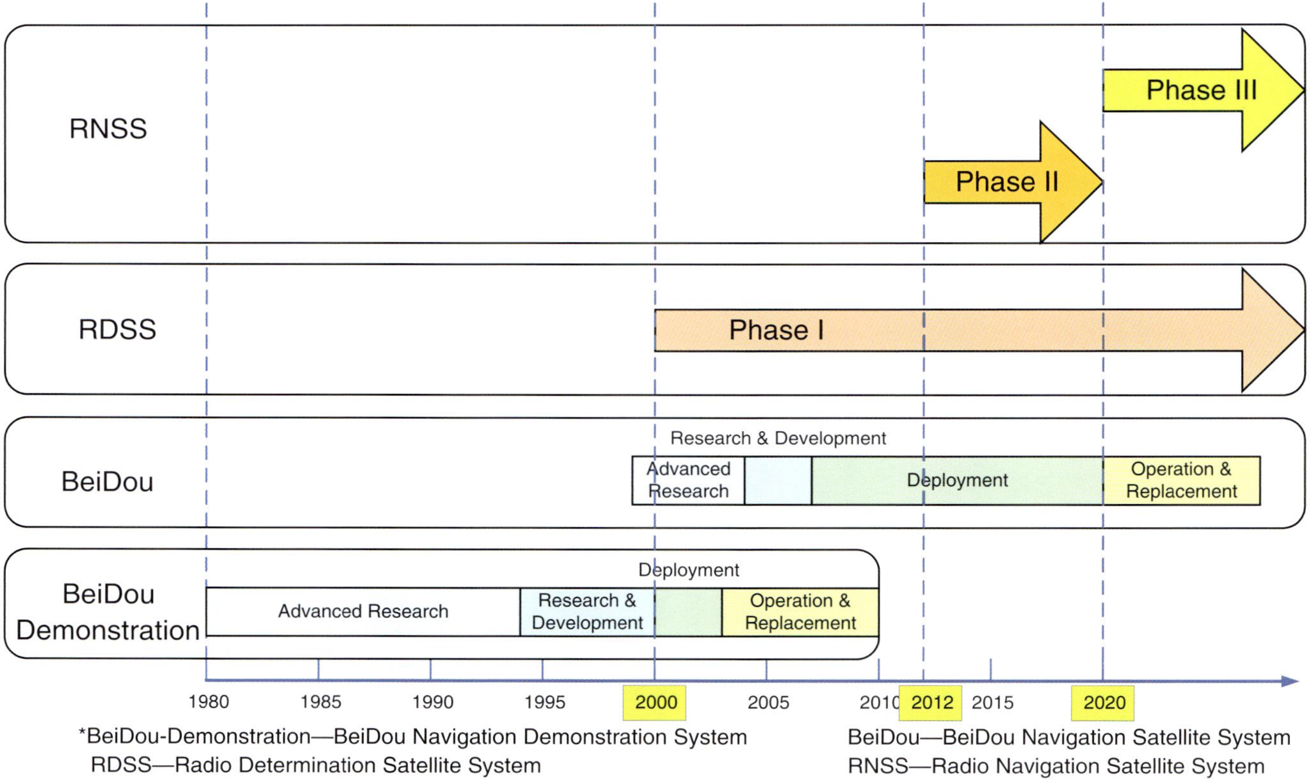

FIGURE 1 Scheme of the BeiDou system.

Time Service Center (NTSC) of the Chinese Academy of Sciences. The coordinate framework of the BeiDou system adopts the China Geodetic Coordinate System 2000 (CGCS2000), which is consistent with the International Terrestrial Reference Frame (ITRF) at the 5 cm level.

Development Status

BeiDou demonstration satellites have been operating for 10 years. In 2010, the Beidou demonstration system was seamlessly transferred to the Beidou navigation system, with its RDSS positioning, timing, and short-message communication services performances improved.

So far, the BeiDou system has made steady progress. From April 2007 to April 2011, eight satellites were launched to form "3+3" basic system except a part used for verification and signal testing and equipment R&D, as is shown by Table 1.

The first medium Earth orbit (MEO) satellite, named COMPASS-M1, was launched on April 14, 2007, and secured the International Telecommunications Union (ITU) related filing. As the first RNSS satellite, many technical experiments have been implemented with it.

The first GEO satellite, named COMPASS-G2, was launched on April 15, 2009. It was used for technical experiments such as GEO orbit determination and time synchronization. Now the satellite is in orbit. The second GEO satellite was launched on January 17, 2010. It was initially located at 160°E to secure the ITU-related filing but, after in-orbit test and experiment verification, it moved to 144.5°E and is now operating well. The third GEO satellite was launched on June 2, 2010. Initially located at 84.6°E to carry out in-orbit tests, it was moved to 84°E and now operates optimally. The fourth GEO satellite was launched on November 1, 2010, and is located at 160°E.

The first IGSO (inclined geosynchronous satellite orbit) satellite was launched on August 1, 2010; the intersection point is at 118°E. The second IGSO satellite was launched December 18, 2010, and the third IGSO satellite was launched on April 10, 2011.

TABLE 1 Launch Record of BeiDou Navigation Satellites

Date	Type	Launch Site	Launch Carrier	Launch Cabin
2007.04.14	MEO	Xichang	CZ-3A	DFH-3
2009.04.15	GEO	Xichang	CZ-3C	DFH-3
2010.01.17	GEO	Xichang	CZ-3C	DFH-3
2010.06.02	GEO	Xichang	CZ-3C	DFH-3
2010.08.01	IGSO	Xichang	CZ-3A	DFH-3
2010.11.01	GEO	Xichang	CZ-3A	DFH-3
2010.12.18	IGSO	Xichang	CZ-3A	DFH-3
2011.04.10	IGSO	Xichang	CZ-3A	DFH-3

In addition, the installment and testing of ground segment components were completed and successfully realized functions. The system is able to provide initial positioning, navigation, and timing services for China and its surrounding areas.

APPLICATION OF BEIDOU NAVIGATION SATELLITE SYSTEM

From Construction to Application

Today, the BeiDou system has played an important role in many fields. It is widely used in transportation, marine fishery, hydrological monitoring, meteorological forecasting, forestry fire prevention, telecommunication and timing, power dispatching, and disaster mitigation and relief.

In disaster prevention and mitigation, the BeiDou system can help to improve rescue response and decision-making capability through broadcasting rapid and timely disaster alerts, scheduling rescue commands, and providing rapid emergency communication. Especially during the Southern China snowstorm and the earthquakes in Wenchuan, Sichuan Province, and Yushu, Qinghai Province, BeiDou terminals sent the disaster and rescue information to a command center in the first place and provided strong support to the search and rescue work.

From Application to Services

Through building an integrated BeiDou application system, the system can integrate national resources through demonstration in various trade sectors and regions and transit from single service into integrative service (Figure 2).

From Regional to Global

The applications of the BeiDou system will gradually extend from China to the Asia-Pacific region, and further extend to global by 2020 (Figure 3).

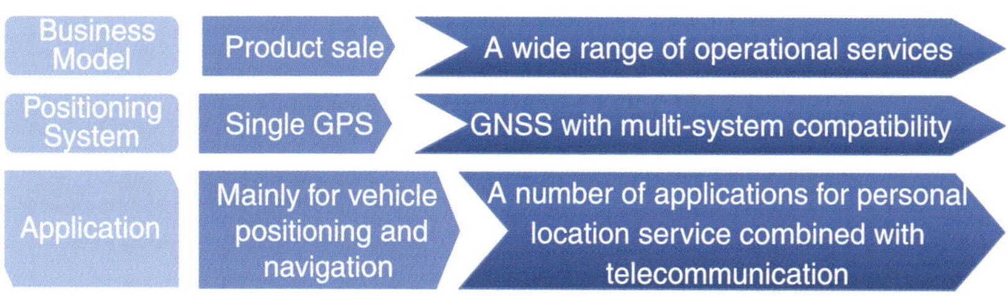

FIGURE 2 The application of BeiDou.

FIGURE 3 The coverage of BeiDou.

INTERNATIONAL COOPERATION

Basic Principle

The BeiDou system adheres to the basic policy of opening up to the outside world and has actively engaged in international space cooperation. It supports all activities that utilize outer space for peaceful purposes and is dedicated to strong cooperation with other countries based on the equality, mutual benefit, and peaceful utilization of outer space and common development.

The international cooperation will be conducted in a phased, focused, nondiscriminatory, and selective approach in accordance with the overall development plan of China's navigation satellite system.

Main Contents

China's international exchange and cooperation in the field of satellite navigation started in the 1990s. In nearly 20 years, various forms of activities have been carried out with extensive results.

In 1994, under the framework of the ITU, China started the BeiDou Navigation Satellite System frequency coordination activities. Satellite network information was submitted in accordance with the BeiDou system's construction plan and progress.

The BeiDou system has actively participated in and supported cooperation in compatibility and interoperability of GNSS, monitoring and evaluation of the service performance of GNSS, the improvement of performance and technology in satellite navigation, GNSS applications, and GNSS international standards of civil aviation and maritime, as well as interference detection and mitigation of GNSS open services.

Main Activities

China has actively participated in bilateral frequency coordination meetings with the United States, the European Union (EU), and Russia. It held several meetings on compatibility and interoperability issues with the EU, attended international academic conferences and forums, such as: the ION annual conference in the United States, the European Navigation Conference, the Munich Satellite Navigation Summit, the World Radiocommunication Conference, and the meetings of ITU study groups and working groups. It also attended ICG-5 in October 2010, and the application for hosting ICG-7 in 2012 gained recognition by UNOOSA (United Nations Office for Outer Space Affairs) and ICG (International Committee on GNSS) members.

CONCLUSIONS

The construction and application of the BeiDou Navigation Satellite System is steadily progressing. The BeiDou system has been attaching great importance to cooperation with other countries to share experience and achievements and will continue to make contributions to GNSS and improving GNSS performance through international cooperation.

The high-level academic workshop will certainly deepen the mutual understanding between the Chinese and American experts in GNSS field. The China Satellite Navigation Office greatly supports the workshop and wishes for fruitful achievements.

U.S. GPS Policy, Programs, and International Cooperation Activities

DAVID A. TURNER
Space and Advanced Technology
U.S. Department of State

OVERVIEW

- U.S. Space-Based Position Navigation and Timing (PNT) Policy
- GPS & Augmentation Programs Status
- International Cooperation Activities

SPACE-BASED PNT GUIDANCE IN U.S. NATIONAL SPACE POLICY

Figure 1 shows the organizational structure for developing GPS policy. The goal is to maintain leadership in the service, provision, and use of GNSS through the following actions:

- Provide civil GPS services free of direct user charges available on a continuous, worldwide basis
- Maintain constellation consistent with published performance standards and interface specifications
- Use foreign PNT services to complement services from GPS
- Encourage global compatibility and interoperability with GPS
- Promote transparency in civil service provision
- Enable market access to industry
- Support international activities to detect and mitigate harmful interference

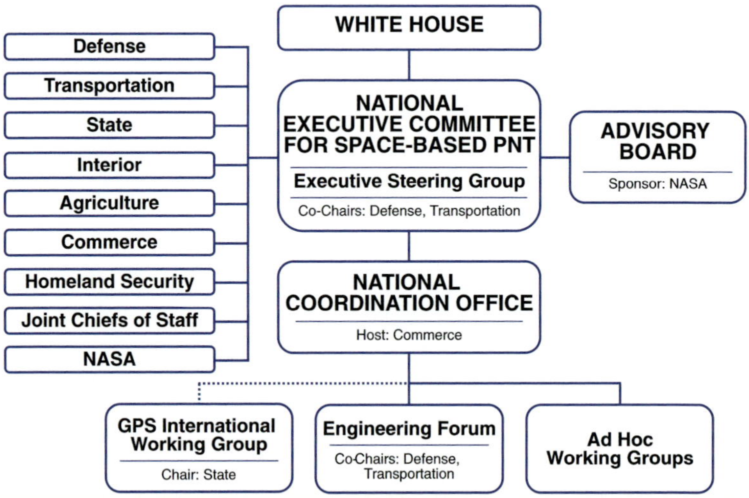

FIGURE 1 U.S. space-based PNT organization structure.

U.S. policy promotes global use of GPS technology:

- No direct user fees for civil GPS services
- Provided on a continuous, worldwide basis
- Open, public signal structures for all civil services
- Promotes equal access for user equipment manufacturing, applications development, and value-added services
- Encourages open, market-driven competition
- Global compatibility and interoperability with GPS
- Service improvements for civil, commercial, and scientific users worldwide
- Protection of radio navigation spectrum from disruption and interference

GPS PROGRAM STATUS

GPS Constellation Status:

- 31 operational satellites (baseline constellation of 24)
  — 11 GPS IIA
  — 12 GPS IIR
  — 7 GPS IIR-M
  — 1 GPS IIF
- 3 additional satellites in residual status
- IIF SV-2 scheduled to launch in July 2011

- IIIA SV-1 scheduled to launch in 2014
- Continuously assessing constellation health to determine launch need
- Global GPS civil service performance commitment met continuously since December 1993 (Figure 2)

GPS Modernization: Increasing Space System Capabilities for Increasing Military/Civil User Benefits:

- 2005–2009—GPS IIR(M): 8 satellites in orbit (Lockheed Martin)
  - — IIA/IIR capabilities plus
  - — Second civil signal (L2C)
  - — M-Code (L1M & L2M)
  - — L5 Demo
  - — Anti-Jam Flex Power
  - — 7.5-year design life
- 2010–2014—GPS IIF: 12 satellites with 2 currently in orbit (Boeing)
  - — IIR(M) capabilities plus
  - — Third civil signal L5
  - — Reprogrammable Nav Processer
  - — Increased accuracy requirement
  - — 12-year design life
- 2014–2024—GPS III: up to 32 satellites (Lockheed Martin)
  - — Increased accuracy
  - — Increased Earth coverage power

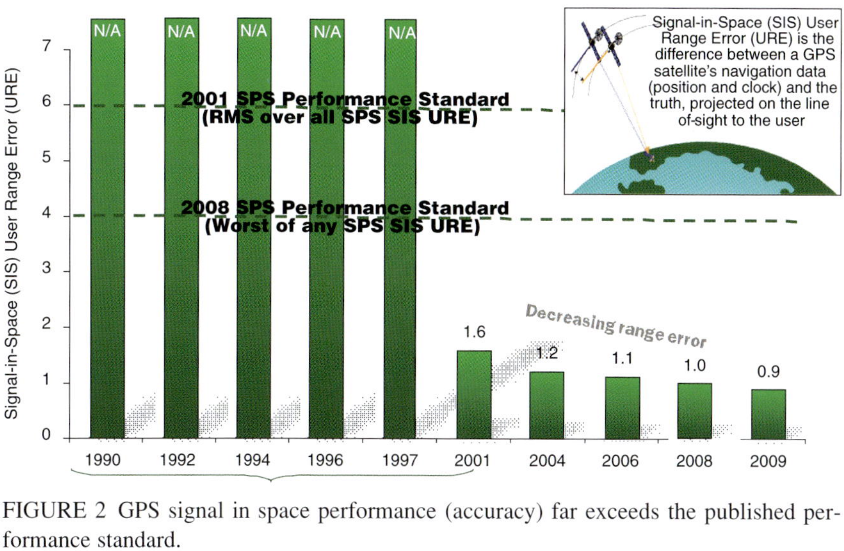

FIGURE 2 GPS signal in space performance (accuracy) far exceeds the published performance standard.

  - — 15-year design life
  - — 4th civil signal (L1C)
  - — Spot Beam for anti jamming (spiral development)
- Operational Control Segment Modernization
  - — Architecture Evolution Plan (AEP)
    - o Transitioned in 2007
    - o Increased worldwide commanding capability
    - o Increased capacity for monitoring of GPS signals
    - o Modern distributed system replaced 1970s mainframes
    - o Current software version (5.5D) enabled SAASM functionality
  - — Next-Generation Control Segment (OCX)
    - o Controls more capable constellation and monitors all GPS signals
    - o \$1.5 billion contract awarded February 25, 2010
    - o Capability delivered incrementally to reduce risk
    - o On track for Preliminary Design Review in 2011
    - o Full capability by ~2016

GPS Modernization will result in the following new civil signals:

- Second civil signal (L2C)
  - — Designed to meet commercial needs
  - — Higher accuracy through ionospheric correction
  - — Available since 2005 without data message
  - — Phased roll-out of CNAV message starting in 2009
  - — Full capability: 24 satellites ~2016
- Third civil signal (L5)
  - — Designed to meet demanding requirements for transportation safety-of-life
  - — Uses highly protected Aeronautical Radio Navigation Service (ARNS) band
  - — 24 satellites ~2018
- Fourth civil signal (L1C)
  - — Designed with international partners for interoperability
  - — Modernized civil signal at L1 frequency
  - — More robust navigation across a broad range of user applications
  - — Improved performance in challenged tracking environments
  - — Original signal retained for backward compatibility
  - — Specification developed in cooperation with industry recently completed
- Launches with GPS III in 2014
- On 24 satellites by ~2021

GPS Modernization: Semi-codeless Transition:

- GPS receivers attain very high accuracy by using "codeless" or "semi-codeless" techniques that exploit the encrypted military GPS signals without actually decoding them. These techniques will no longer be necessary once the new civil GPS signals are fully operational.
- The U.S. government published a notice for users to transition to GPS civil-coded signals by December 31, 2020, based on the current launch schedule and projected budget. This will provide time for an orderly and systematic transition.
- The U.S. government led community-wide collaboration on this transition plan.
- The United States is committed to continually improving GPS services as users complete a timely transition to dual-coded civil GPS equipment.

AUGMENTATION PROGRAMS STATUS

Wide Area Augmentation System—Phased Upgrades:

- Phase I: Initial Operational Capability (July 2003) Completed
  —Provided lateral navigation/vertical navigation/limited localizer performance with vertical guidance (LNAV/VNAV/LPV) capability
- Phase II: Full LPV (FLP) (2003–2008) Completed
  —Improved LPV availability in Continental United States (CONUS) and Alaska
  —Expanded WAAS coverage to Mexico and Canada
- Phase III: Full LPV-200 Performance (2009–2013)
  —Software enhancements, hardware upgrades
  —Steady-state operations and maintenance
  —Transition to FAA performed second level engineering support
  —Begin GPS L5 transition activities
- Phase IV: Dual Frequency (L1,L5) Operations (2013–2028)
  —Complete GPS L5 transition
  —Will significantly improve availability and continuity during severe solar activity
  —Provide additional protection against GPS interference
  —Will continue to support single frequency users

Nationwide Differential GPS System (Figure 3):

- Expansion of maritime differential GPS (DGPS) network to cover terrestrial United States
- Built to international standard adopted in 50+ countries

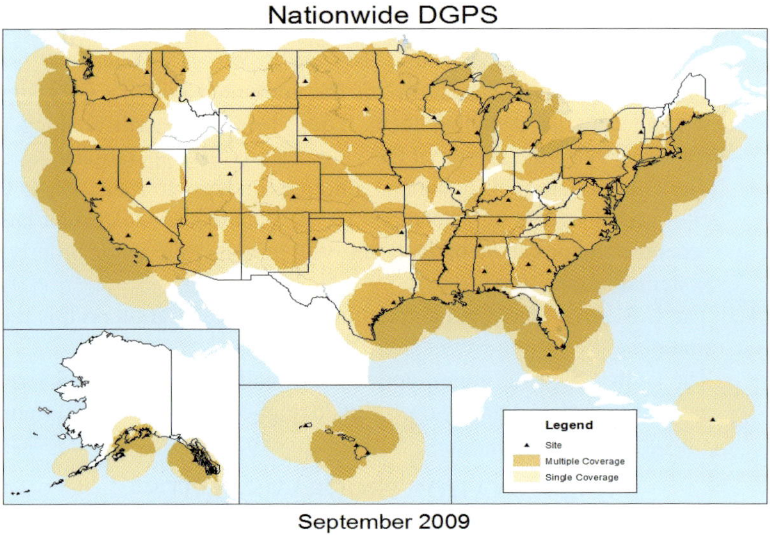

FIGURE 3 Nationwide DGPS network.

- Operated/managed by U.S. Coast Guard as a Combined NDGPS (Maritime + Department of Transportation sites + ACOE sites)
- System Specifications
  - —Corrections broadcast at 285 and 325 kHz using minimum shift keying (MSK) modulation
  - —Real-time differential GPS corrections provided in Radio Technical Commission for Maritime Services (RTCM) SC-104 format
  - —No data encryption
  - —Real-time differential corrections for mobile and static applications
  - —Single coverage terrestrial over 92 percent of CONUS; double coverage over 65 percent of CONUS

National Continuously Operating Reference Stations (CORS) (Figure 4):

- Enables highly accurate, 3-D positioning
  - —Centimeter-level precision
  - —Tied to National Spatial Reference System
- 1,200+ sites operated by 200+ public, private, academic organizations
- National Oceanographic and Atmospheric Administration's (NOAA) Online Positioning User Service (OPUS) automatically processes coordinates submitted via the web from around the world

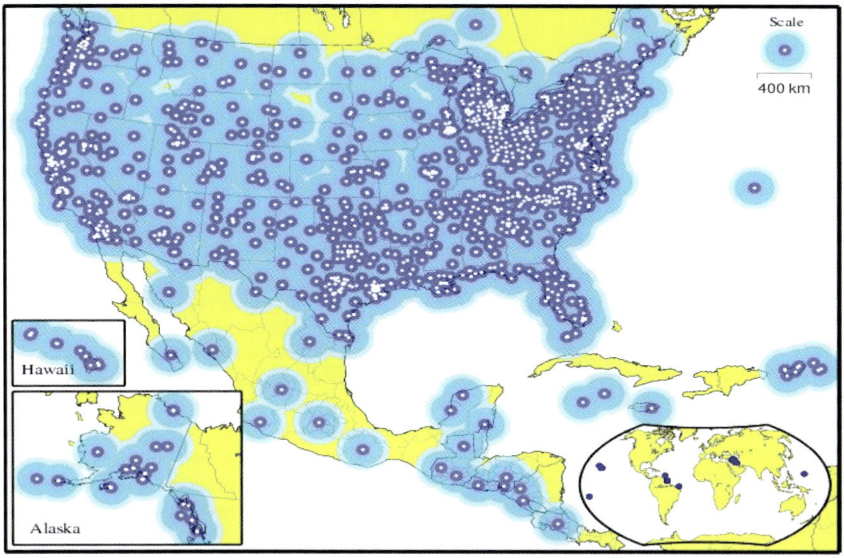

FIGURE 4 CORS network.

- OPUS-RS (Rapid Static) declared operational in 2007
- NOAA considering support for real-time networks

INTERNATIONAL COOPERATION ACTIVITIES

Excerpts from U.S. Space-Based PNT Policy focused on international relations:

- Goals include maintaining U.S. space-based PNT systems and services as essential components of internationally accepted PNT services, and promoting U.S. technological leadership in applications involving space-based PNT services.
- To achieve this, the U.S. government shall: encourage foreign development of PNT services/systems based on GPS; seek to ensure foreign space-based PNT systems are interoperable with civil GPS and augmentations; and, at a minimum, ensure compatibility.
- The Secretary of State shall promote the use of civil aspects of GPS and its augmentation services and standards with foreign governments and other international organizations, and lead negotiations with foreign governments and international organizations regarding civil PNT matters.

U.S. objectives in working with other GNSS service providers include:

- Ensuring compatibility, defined as the ability of U.S. and non-U.S. space-based PNT services to be used separately or together without interfering with each individual service or signal, involving both radio frequency compatibility and spectral separation between M-code and other signals.
- Achieving interoperability, defined as the ability of civil U.S. and non-U.S. space-based PNT services to be used together to provide the user better capabilities than would be achieved by relying solely on one service or signal, with the primary focus on the common L1C and L5 signals.
- Ensure a level playing field in the global marketplace.

U.S. GNSS international objectives are pursued through bilateral and multilateral cooperation as follows:

- U.S.-EU GPS-Galileo Cooperation Agreement signed in June 2004
  - Four working groups set up under the Agreement
- U.S.-Japan Joint Statement on GPS Cooperation 1998
  - Quasi-Zenith Satellite System (QZSS) designed to be fully compatible and highly interoperable with GPS
  - Bilateral agreements to set up QZSS monitoring stations in Hawaii and Guam
- U.S.-Russia Joint Statement issued December 2004
  - Working Groups on navigation signal compatibility and interoperability, and search and rescue service
- U.S.-China operator-to-operator coordination meetings held under ITU auspices in:
  - Geneva, Switzerland - June 2007
  - Xian, China - May 2008
  - Geneva, Switzerland - October 2008
  - Hainan, China - December 2009
  - Coordination completed in Chengdu, China, in September 2010
  - The United States is interested in engaging in further bilateral discussions with China on civil GNSS services and applications

- U.S.-India Joint Statement on GNSS Cooperation 2007
  - Technical Meetings focused on GPS-India Regional Navigation Satellite System (IRNSS) compatibility and interoperability held in 2008 and 2009
  - Continuation of ITU compatibility coordination is pending

- U.S.-Australia Joint Delegation Statement on Cooperation in the Civil Use of GPS in 2007
  - — Bilateral meeting in Washington, D.C., October 26–27, 2010
  - — GNSS and applications to be included in expanded space cooperation, as discussed in an October 27 Joint Announcement

International Committee on GNSS:

- Emerged from Third UN Conference on the Exploration and Peaceful Uses of Outer Space July 1999
  - — Promote the use of GNSS and its integration into infrastructures, particularly in developing countries
  - — Encourage compatibility and interoperability among global and regional systems
- Members include:
  - — GNSS providers (United States, European Union, Russia, China, India, Japan)
  - — Other Member States of the United Nations
  - — International organizations/associations
- Providers Forum
  - — Six space segment providers are members
  - — Purpose:
    - o Focused discussions on compatibility and interoperability, encouraging development of complementary systems
    - o Exchange detailed information on systems and service provision plans
    - o Exchange views on ICG work plan and activities
  - — Providers have agreed that all GNSS signals and services must be compatible, and open signals and services should also be interoperable to the maximum extent possible
  - — Working definition of compatibility includes respect for spectral separation between each system's authorized service signals and other systems' signals
  - — Interoperability definition addresses signal, geodetic reference frame realization, and system time steerage considerations

APEC (Asia-Pacific Economic Cooperation) GNSS Implementation Team

- GNSS Implementation Team (GIT) established in 2002
- Mission—Promote implementation of regional GNSS augmentation systems to enhance inter-modal transportation and recommend actions to be considered in the Asia Pacific Region

- Reports to Transportation Working Group through the Inter-modal Experts Group
- Adopted a GNSS Strategy designed to promote adoption of GNSS technologies throughout the Asia Pacific region, especially with regard to transportation

CONCLUSION

- GPS performance is better than ever and will continue to improve
  — Augmentations enable even higher performance
  — New civil GPS signal available now
  — Many additional upgrades scheduled
- U.S. policy encourages worldwide use of civil GPS and augmentations
- International cooperation is a priority
- In pursuit of systems compatible and interoperable with GPS

Research Report on GNSS Interoperability

LU XIAOCHUN
National Time Service Center, Chinese Academy of Sciences

LU JUN
Beijing Institute of Tracking and Telecommunications Technology

BAI YAN, HAN TAO, and WANG XUE
National Time Service Center, Chinese Academy of Science

INTRODUCTION

Interoperability has become a focus of Global Navigation Satellite Systems (GNSS) and a development aspect, and incurs much focus among the world. For the purpose of maximum benefit, a series of interoperability researches and cooperation are put forward.

Research of interoperability includes both technical factors, such as signal design, satellite payload, and user terminal, and nontechnical factors, such as market and industry. One should consider not only combining with other systems, but also vindicating one's own benefits and maintaining some independence.

Thus, research of interoperability should be approached with consideration of both technical factors and nontechnical factors.

INTEROPERABILITY DEFINITION PARSING

Elements of interoperability include coordinate reference frame, time reference frame, open signal, and the constellation. The resources probably applied in interoperability are global navigation satellite systems and their services, regional navigation satellite systems and their services, and augmentation systems and their services. Finally, the purpose of interoperability is to provide better services at the user level, including higher accuracy, better reliability, better cost-effectiveness, and more user satisfaction (different types of users, different application areas).

DISCUSSION OF INTEROPERABILITY PHASES BETWEEN BEIDOU GLOBAL SYSTEM AND GPS

Actuality of BeiDou and GPS

Coordinate Reference Frame

BeiDou uses CGS2000 as its coordinate reference frame and keeps a centimeter level bias with ITRF. GPS uses WGS84 as its coordinate reference frame and keeps a centimeter level bias with ITRF.

Time Reference Frame

The time reference frame of BeiDou is called "BDT"; the difference between BDT and UTC (NTSC) is less than 50 ns. The time reference frame of GPS is called "GPST"; the difference between GPST and UTC (USNO) is less than 28 ns.

Open Signals

The civil signal in the GPS L1 band is C/A code-BPSK(1); future civil signals in the L1 band are MBOC (TMBOC in pilot channel and BOC(1,1) in data channel).

The civil signal in the second phase of BeiDou is BPSK(2), and it provides service in the important area (30E~180E, 70S~70N); the civil signal in the third phase is MBOC (TMBOC in pilot channel and BOC(1,1) in data channel) (Tables 1 and 2).

TABLE 1 Signal in the Second and Third Phases of BeiDou

Phase	Signal	Center frequency (MHz)	Code rate (MHz)	Modulation	Services
BeiDou 2nd.	B1(I)	1561.098	2.046	BPSK	Open
BeiDou 3rd.	B1x/y	1575.42	1.023	TMBOC(6,1,4/33)+BOC(1,1)	Open

TABLE 2 Signal of Current GPS and Future GPS

Phase	Signal	Center frequency (MHz)	Code rate (MHz)	Modulation	Services
Current GPS	L1C/A	1575.42	1.023	BPSK	Open
Future GPS	L1 C	1575.42	1.023	TMBOC(6,1,4/33)+BOC(1,1)	Open

Currently the BeiDou system has launched eight satellites of phase-2 and elementary form of positioning ability. Domestic industries have designed CMOS chip in L1 band (1561.098 MHz and 1575.42 MHz), and they plan to produce receivers able to receive GPS L1 C/A and BeiDou B1 BPSK(2), which make base of interoperability.

Constellation

GPS now has 30 satellites in orbit. The BeiDou regional system is designed to have 5 geostationary (GEO), 3 inclined geosynchronous orbit (IGSO) satellites, and 4 medium Earth orbit (MEO) satellites (Figure 1); while the BeiDou global system is designed to have 5 GEO, 3 IGSO, and 24 MEO satellites (Figure 2).

Suggestion of Interoperability Between BeiDou and GPS

According to current resources, interoperability between BeiDou B1 and GPS L1 should be staged in three phases (Figure 3):

Phase 1: From 2012 until GPS TMBOC operates.
Phase 2: From GPS TMBOC operation to BeiDou TMBOC functions (2020).
Phase 3: After BeiDou TMBOC operates (2020).

Phase 1

Interoperability between GPS L1 BPSK(1) in 1575.42 MHz and BeiDou regional system B1 BPSK(2) in 1561.098 MHz (Figure 4).

Phase 2

Interoperability between GPS L1 TMBOC in 1575.42 MHz and BeiDou regional system B1 BPSK(2) in 1561.098 MHz (Figure 5).

Phase 3

Interoperability between GPS L1-TMBOC in 1575.42 MHz and BeiDou global system B1-TMBOC in 1575.42 MHz (Figure 6).

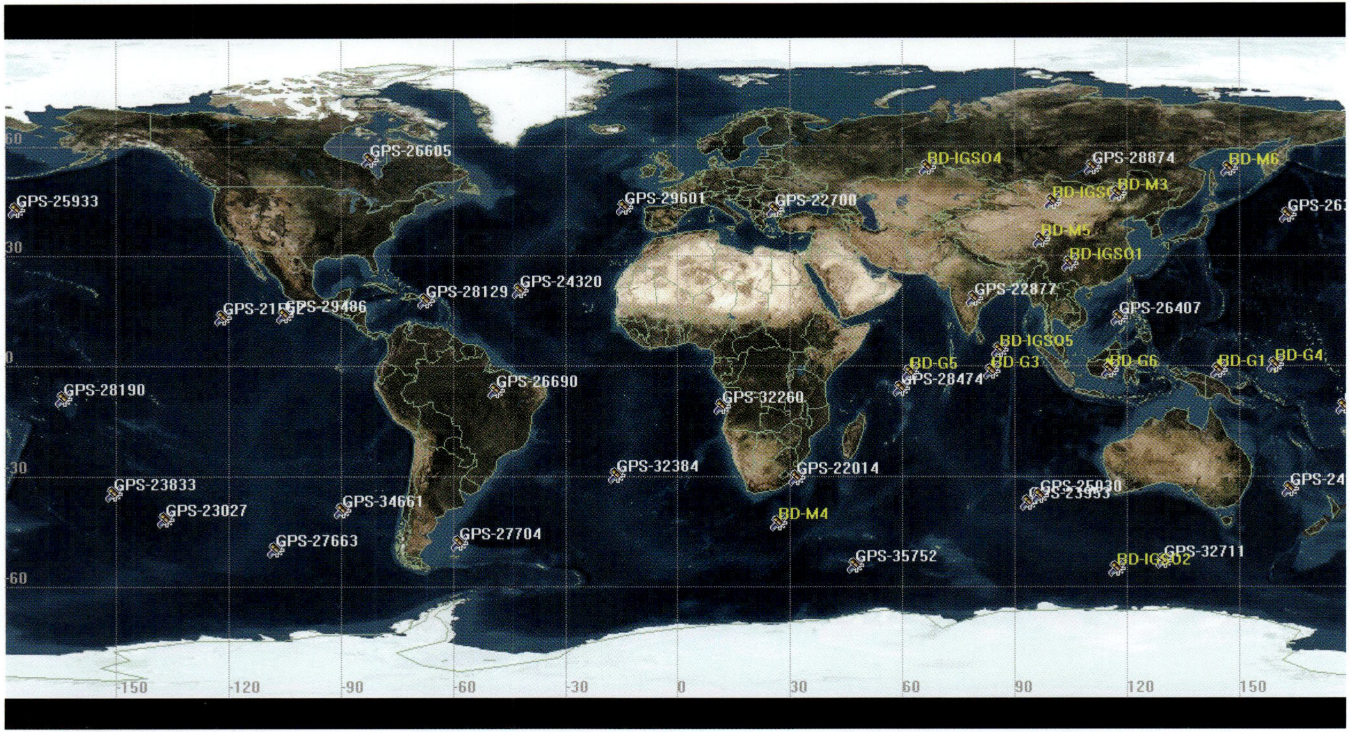

FIGURE 1 Constellation of GPS and BeiDou regional system.

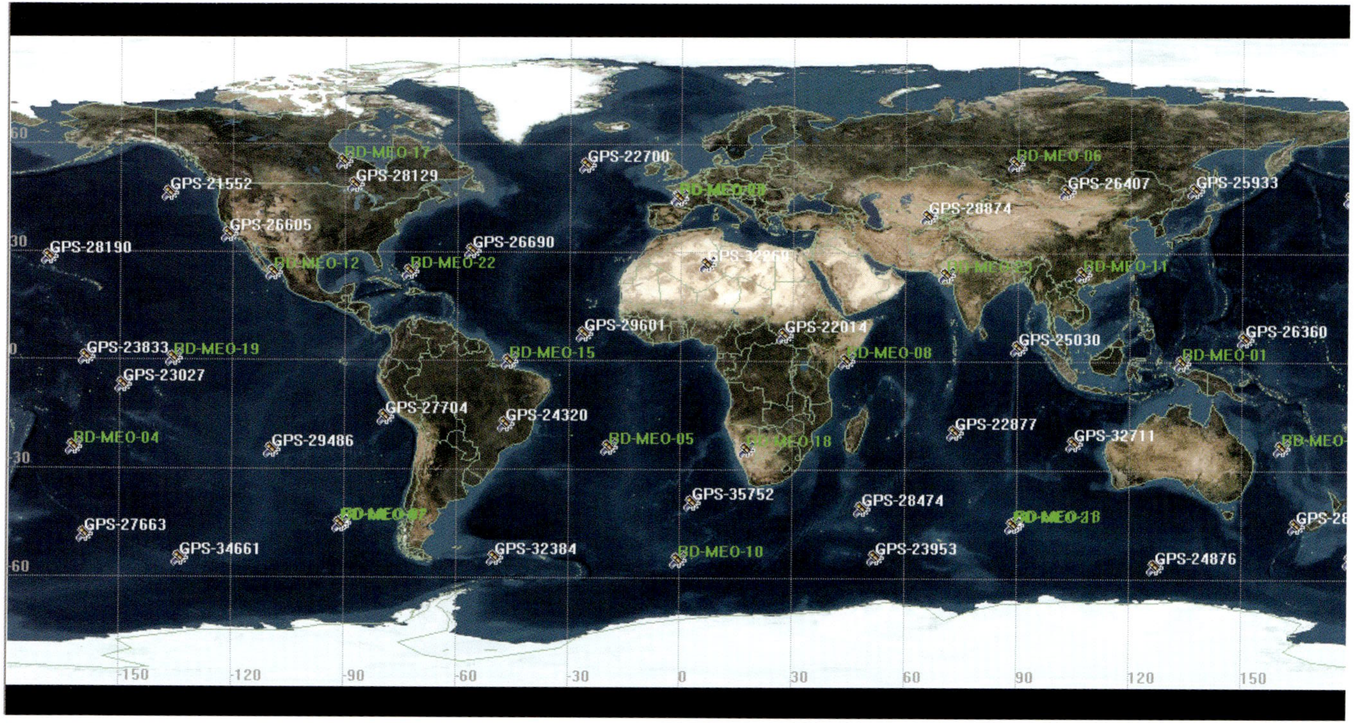

FIGURE 2 Constellation of GPS and BeiDou global system.

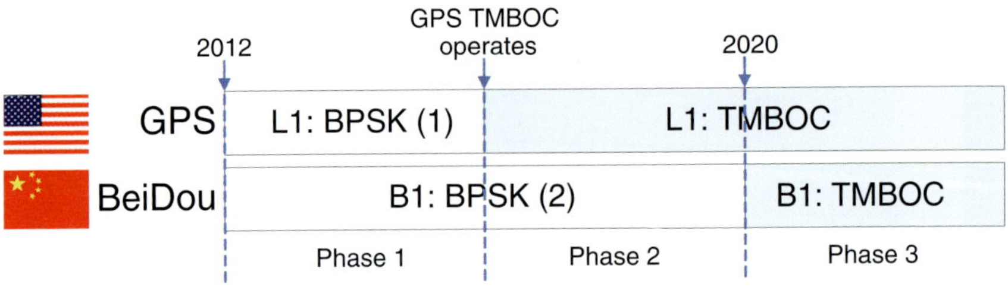

FIGURE 3 Phases of interoperability between BeiDou and GPS.

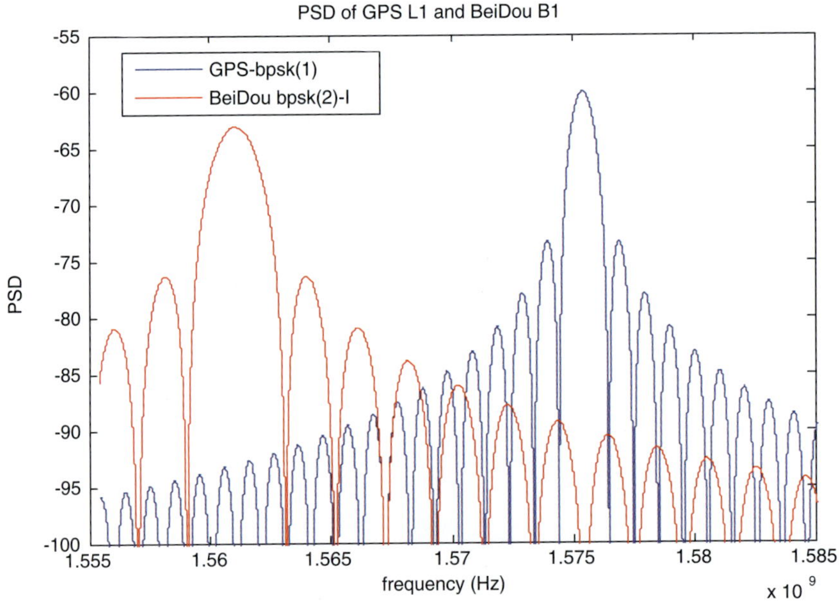

FIGURE 4 Frequency spectrum in interoperability phase 1.

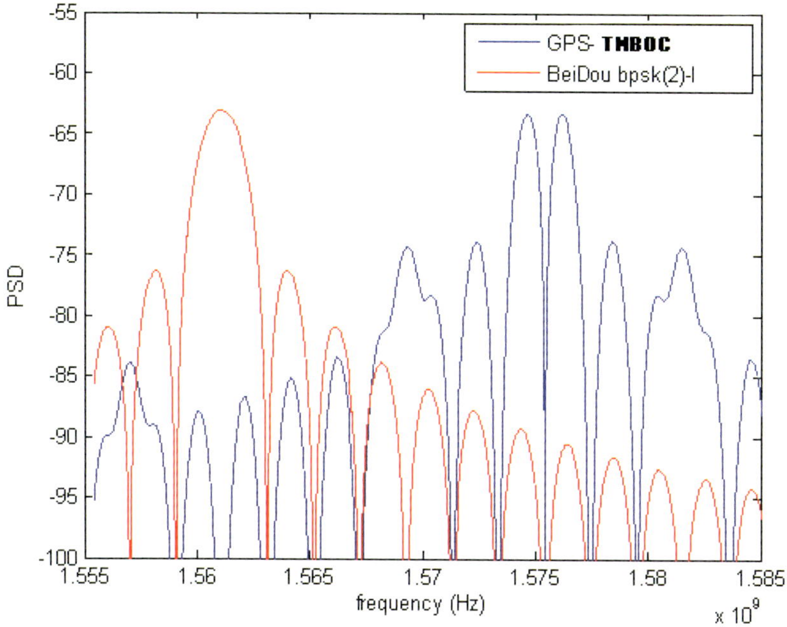

FIGURE 5 Frequency spectrum in interoperability phase 2.

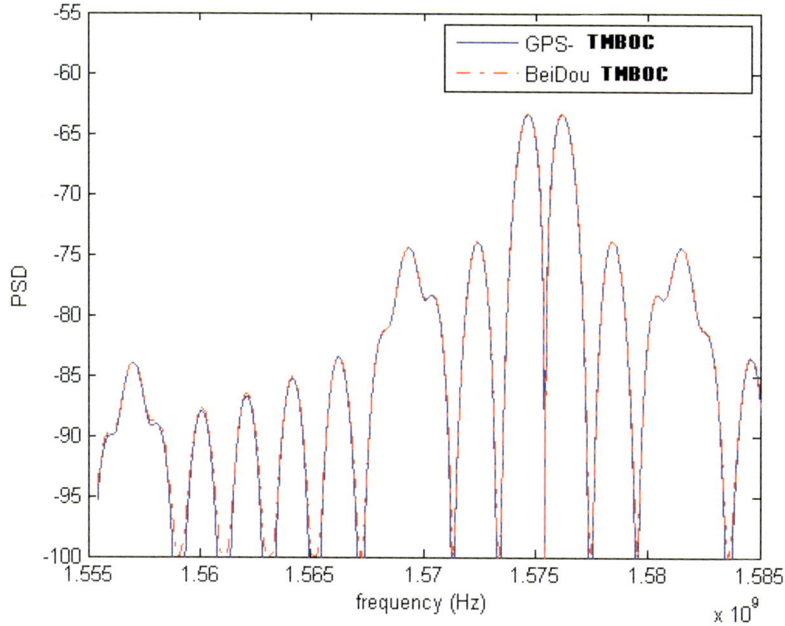

FIGURE 6 Frequency spectrum in interoperability phase 3.

FEASIBILITY ANALYSIS OF INTEROPERABILITY BETWEEN BEIDOU B1 BPSK(2) AND GPS BPSK(1)

Signal Performance Analysis

Correlation peak

Correlation peak can be calculated as:

$$CCF(\varepsilon)=\frac{\int_0^{T_p} S_{BB-\mathrm{PreProc}}(t)\cdot S^*_{\mathrm{Ref}}(t-\varepsilon)\,dt}{\sqrt{\left(\int_0^{T_p}\left|S_{BB-\mathrm{PreProc}}(t)\right|^2 dt\right)\cdot\left(\int_0^{T_p}\left|S_{\mathrm{Ref}}(t)\right|^2 dt\right)}} \tag{4.1}$$

where $S_{BB-\mathrm{PreProc}}$ is the base band signal (after pretreatment); reference signal S_{Ref} is the copy of ideally base band signal from receiver; integral time T_p corresponds to the main period of reference signal. The correlation peak is correlated to ranging accuracy, anti-multipath, and anti-jamming of a signal. The sharper the peak is, the better performance a signal has. From Figure 7, BPSK(2) has a sharper correlation peak than BPSK(1).

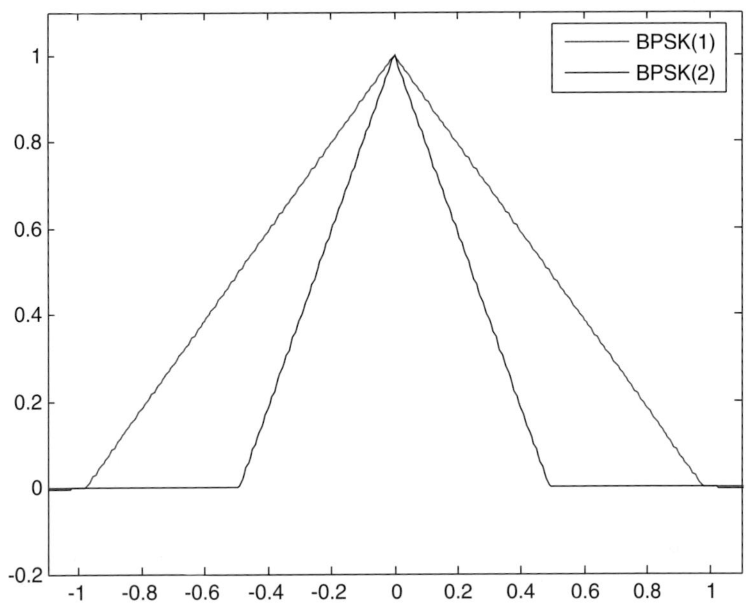

FIGURE 7 Correlation peak of BPSK(1) and BPSK(2).

Gabor Bandwidth

Gabor bandwidth is the best index to estimate the connection from receive-bandwidth to tracking accuracy; the greater the Gabor bandwidth is, the better the signal's tracking accuracy. Under the same code loop bandwidth and same receive carrier to noise ratio, the root mean square (RMS) code tracking accuracy depends on RMS bandwidth.

$$\beta_{rms} = \sqrt{\int_{-\frac{\beta_r}{2}}^{\frac{\beta_r}{2}} f^2 G_S(f)\,df} \tag{4.2}$$

where β_{rms} is RMS bandwidth (namely Gabor bandwidth), β_r is bilateral receive bandwidth, and $G_S(f)$ is PSD.

The RMS bandwidth of different modulations is shown in Table 3. The anti-jamming of BPSK(2) is better than BPSK(1), but weaker than MBOC.

TABLE 3 Gabor Bandwidth

Modulation	BPSK(1)	BPSK(2)	CBOC+	CBOC-	TMBoc	BOCsin(2,2)
Gabor bandwidth	1.4415×10^5	2.0525×10^5	3.0990×10^5	3.5348×10^5	3.552×10^5	3.5558×10^5

Tracking Error

$$\sigma^2_{NELP} = \frac{B_L(1-0.25B_LT)\int_{-\beta_r/2}^{\beta_r/2} G_S(f)\sin^2(\pi f\Delta)\,df}{\frac{C}{N_0}\left(2\pi\int_{-\beta_r/2}^{\beta_r/2} fG_S(f)\sin(\pi f\Delta)\,df\right)^2} \times \left[1+\frac{\int_{-\beta_r/2}^{\beta_r/2} G_S(f)\cos^2(\pi f\Delta)\,df}{T\frac{C}{N_0}\left(\int_{-\beta_r/2}^{\beta_r/2} G_S(f)\cos(\pi f\Delta)\,df\right)^2}\right] \tag{4.3}$$

Suppose the interference signal power spectrum is white, one-side power spectrum is N_0, signal-receive power is P_c, and forward bandwidth is β_r. The bilateral power spectrum of interference signal baseband equivalent $w(t)$ is twice the unilateral power spectrum of radio interference; power (P_s) of available signal envelope $s(t)$ is twice of radio signal, influence from the channel is equal to baseband transmit function $H(f)$, $G_S(f)$ is signal power spectrum, T is integral time, and Δ is correlator pace (unit: second) (Table 4).

TABLE 4 Simulation Parameters

Delt = 5	IntTime = 0.004 s	Dllw = 0.5Hz	d = 1/(12*1.023*10^6)
Space of power spectrum frequency	Integral time	Loop bandwidth	Correlator space

Figure 8 shows that BPSK(2) has better tracking error than BPSK(1).

Multipath

Receive signal with multipath can be equated to:

$$r(t) = a_0 e^{j\phi_0} x(t - \tau_0) + \sum_{n=1}^{N} a_n e^{j\phi_n} x(t - \tau_n) \quad (4.4)$$

where a_0 is the extent of firsthand signal; ϕ_0 is the phase of firsthand signal; $x(t)$ is the complex envelope of sending signal; τ_0 is the time delay of firsthand signal; N is the number of path of the multipath signal; a_n is the extent of multipath signal; ϕ_n is the phase of the multipath signal; and τ_n is the time delay of the multipath signal.

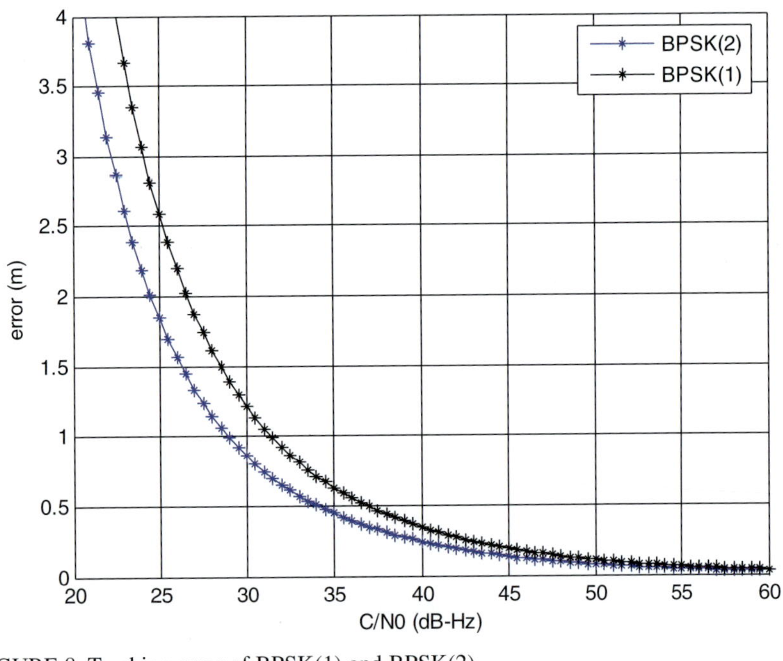

FIGURE 8 Tracking error of BPSK(1) and BPSK(2).

Multipath error can be depicted as:

$$\varepsilon_\tau \approx -\frac{\pm 2a_1\int_{-\beta_r/2}^{\beta_r/2} S(f)\sin(-2\pi f\tilde{\tau}_1)\sin(\pi fd)df}{4\pi a_0\int_{-\beta_r/2}^{\beta_r/2} fS(f)\sin(\pi fd)df \pm 4\pi a_1\int_{-\beta_r/2}^{\beta_r/2} fS(f)\sin(-2\pi f\tilde{\tau}_1)\sin(\pi fd)df}$$

$$\approx \frac{\pm\tilde{a}_1\int_{-\beta_r/2}^{\beta_r/2} S(f)\sin(2\pi f\tilde{\tau}_1)\sin(\pi fd)df}{2\pi\int_{-\beta_r/2}^{\beta_r/2} fS(f)\sin(\pi fd)\left[1\pm\tilde{a}_1\cos(2\pi f\tilde{\tau}_1)\right]df} \quad (4.5)$$

where $\tilde{a}_1 = a_1 / a_0$ is the extent ratio from the multipath signal to the firsthand signal, β_r is the front bandwidth of receiver, and d is the correlator space.

The mean multipath error $A(\tau)$ can be calculated by:

$$A(\tau) = \frac{1}{\tau}\int_0^\tau \left|\max(E(x)) - \min(E(x))\right| dx \quad (4.6)$$

where $E(x)$ is the curve function of multipath error envelope, and τ is code time delay. Then the even multipath error is:

$$\varepsilon_a(\tau_1) = \frac{1}{\tau_1}\int_0^{\tau_1}\left[\frac{\left|\varepsilon(\tau)\right|_{\varphi=0}\left|+\right|\varepsilon(\tau)\left|_{\varphi=180}\right|}{2}\right]d\tau \quad (4.7)$$

where $\varepsilon(\tau)$ is the function of multipath error envelope, $\varepsilon_a(\tau_1)$ is the function of even multipath error, τ_1 and τ are multipath signal time delay (Table 5, Figures 9 and 10).

TABLE 5 Simulation Parameters

a1 = -5 dB	B = 30 MHz	d = 1/20 code
Extent ratio from multipath signal to firsthand signal	Front receive bandwidth	Correlator space

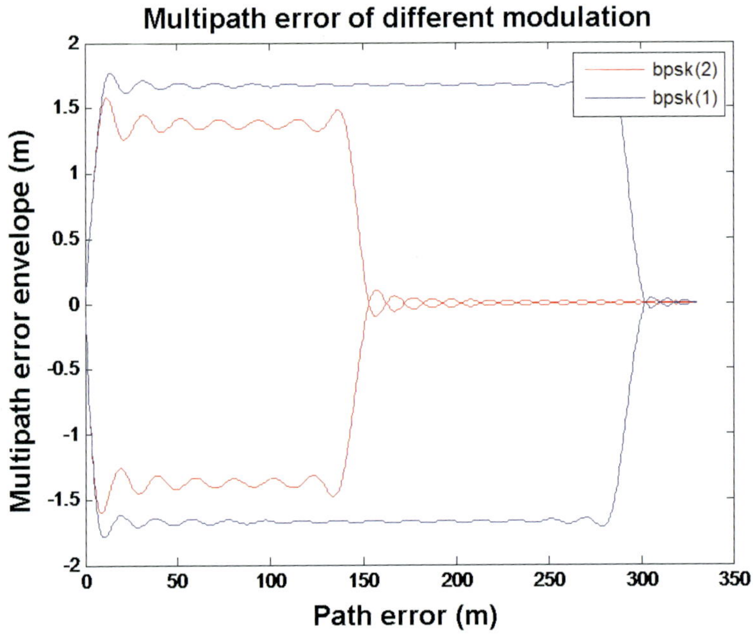

FIGURE 9 Multipath error of BPSK(1) and BPSK(2).

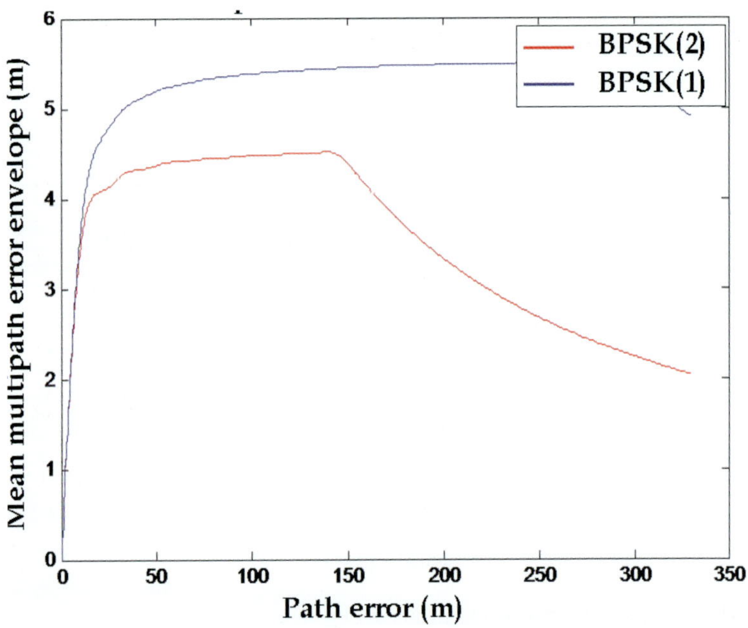

FIGURE 10 Mean multipath error of BPSK(1) and BPSK(2).

Code Cross-correlation

Code cross-correlation can validate the cross-correlation between BeiDou ranging code and GPS L1 C/A code. Simulate the maximum value after uniformed correlation between 32 GPS codes to 32 GPS codes, 13 BeiDou codes to 13 BeiDou codes, and 32 GPS codes to 13 BeiDou codes, as shown in Figures 11–13. The colors in these figures represent the maximum correlation values.

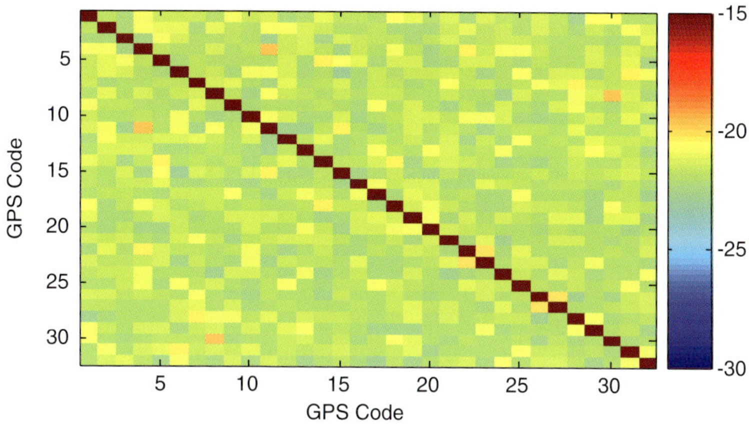

FIGURE 11 Cross-correlation of GPS ranging code.

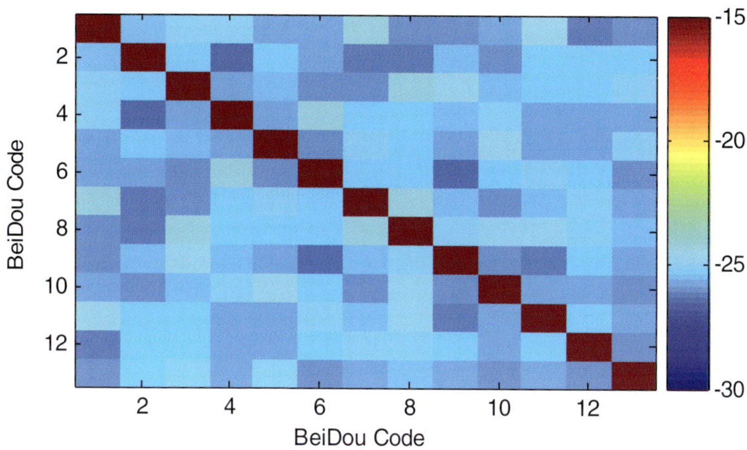

FIGURE 12 Cross-correlation of BeiDou ranging code.

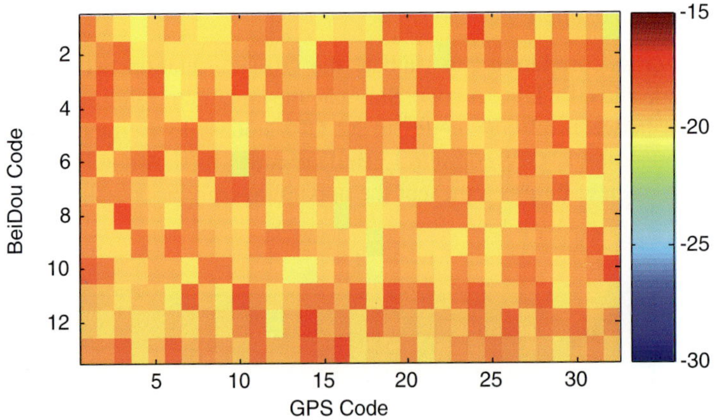

FIGURE 13 Cross-correlation of GPS ranging code and BeiDou ranging code.

Brief Summary

BeiDou B1 BPSK(2) has a better tracking performance, better anti-jamming performance, and better anti-multipath performance than GPS L1 BPSK(1). Interoperability of the two signals makes a higher receiver ranging accuracy.

Interoperability at different frequencies enhances the anti-jamming and signal reliability.

Service Performance

Accuracy

Service accuracy is employed to depict the difference between the real value and measurement value of positioning, timing, and velocity measure. Space signal accuracy includes user range error (URE), user range ratio error (URRE), user range acceleration error (URAE), and user timing error.

The positioning error depends on pseudo range measurement noise, satellite location error, and positioning dilution of precision (PDOP).

$$\sigma_u = PDOP\sigma \tag{4.8}$$

where PDOP is a nonrandom factor determined by the location of the user and satellites; σ is related to URE and is a random factor.

Service accuracy is the difference between the users' real position, velocity, and timing value and their measurements. Accuracy = UERE × PDOP, where

UERE is made up of URE and UEE $\left(UERE = \sqrt{(URE)^2 + (UEE)^2}\right)$; UERE is related with satellite clock error, ephemeris error, atmosphere model error, multipath error, and receiver clock error. In a given system, with the same URE and same UEE, the positioning accuracy is determined by PDOP.

Availability

Service availability is the time ratio of congruous threshold value of positioning, timing, and velocity measurement in a prescriptive time period (usually a recursive period) and a given region.

Suppose the threshold of positioning accuracy and timing accuracy are σ_p, σ_t. The URE is related to positioning accuracy:

$$\sigma_p = PDOP \cdot URE \tag{4.9}$$

Timing accuracy is (GPS SPS 2008):

$$\sigma_t = TDOP \cdot \Delta t = TDOP \times URE \div c\,(c \text{ is velocity of light}) \tag{4.10}$$

Thus, the threshold of positioning accuracy and timing accuracy in a given system is:

$$\begin{pmatrix} \sigma_P \\ \sigma_t \end{pmatrix} = \begin{pmatrix} PDOP \\ TDOP \div c \end{pmatrix} \times URE \tag{4.11}$$

For the same threshold of positioning accuracy and timing accuracy, decreases of PDOP and TDOP will help to add to the threshold. Because the range of URE will be enlarged, which leads to the enhancement of GNSS availability, this extension is related with the distribution of URE.

Chinese Area

Calculate the PDOP and TDOP, shown as Tables 6 and 7.

Calculate PDOP and TDOP of interoperability constellation, GPS constellation, and BeiDou constellation in seven days (a return period) point by point both in the region of China. Then compare their average to sole system, as follows (Figures 14–17).

TABLE 6 PDOP Value of GPS+BeiDou Regional System

	Place	GPS	BeiDou	GPS+BeiDou	Place	GPS	BeiDou	GPS+BeiDou
Mean		1.697	2.360	1.224		1.666	2.422	1.193
Max	Lintong	2.605	3.162	1.689	Beijing	2.662	3.250	1.683
Min		1.297	1.537	0.973		1.205	1.551	0.944
Mean		1.786	5.036	1.376		1.633	2.153	1.142
Max	Kashi	5.660	12.611	2.055	Sanya	2.555	2.809	1.602
Min		1.243	1.943	1.045		1.108	1.460	0.833
Mean		1.619	3.528	1.189		1.646	3.445	1.245
Max	Mohe	2.493	5.522	1.524	Tokyo	2.455	5.315	1.746
Min		1.202	1.873	0.919		1.251	1.733	1.014
Mean		1.522	2.107	1.085				
Max	Singapore	2.018	2.748	1.295				
Min		1.253	1.447	0.905				

TABLE 7 TDOP Value of GPS+BeiDou Regional System

	Place	GPS	BeiDou	GPS+BeiDou	Place	GPS	BeiDou	GPS+BeiDou
Mean		0.884	1.433	0.661		0.865	1.481	0.630
Max	Lintong	1.643	1.965	1.084	Beijing	1.635	2.049	1.057
Min		0.518	0.788	0.451		0.494	0.717	0.449
Mean		0.918	3.465	0.721		0.824	1.353	0.622
Max	Kashi	3.714	8.690	1.361	Sanya	1.446	1.855	0.969
Min		0.526	0.875	0.481		0.465	0.790	0.409
Mean		0.806	2.458	0.589		0.847	2.435	0.676
Max	Mohe	1.574	4.369	0.824	Tokyo	1.479	4.123	1.052
Min		0.506	0.834	0.409		0.577	0.887	0.493
Mean		0.721	1.318	0.579				
Max	Singapore	1.064	1.786	0.767				
Min		0.534	0.813	0.460				

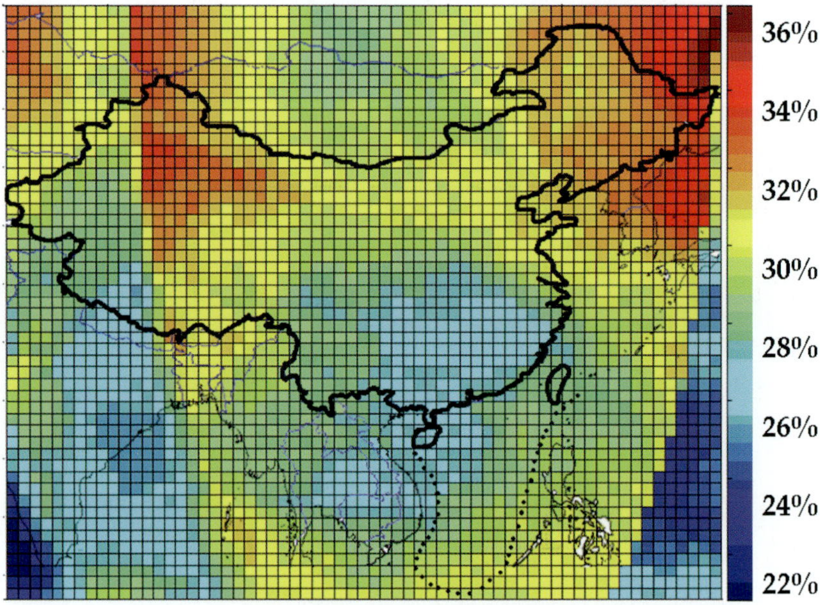

FIGURE 14 Average PDOP reducing in China (compare with GPS).

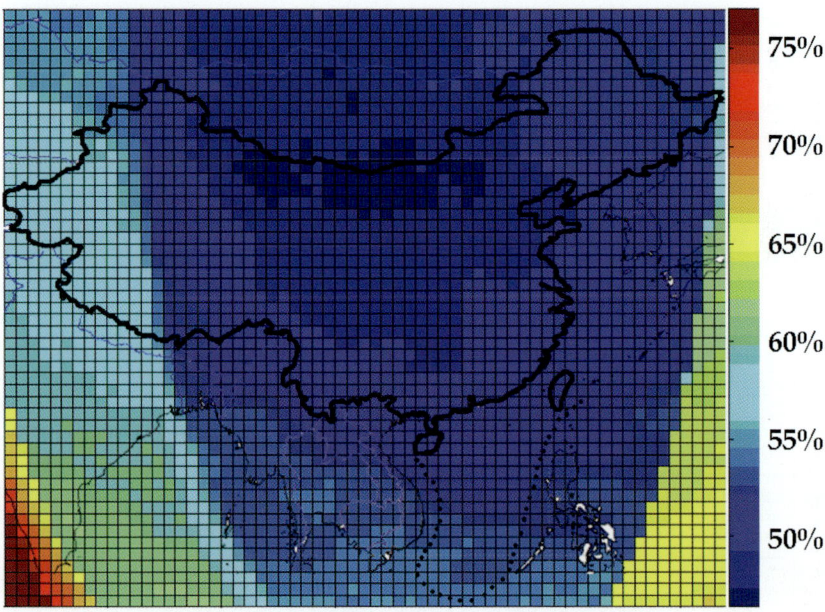

FIGURE 15 Average PDOP reducing in China (compare with BeiDou).

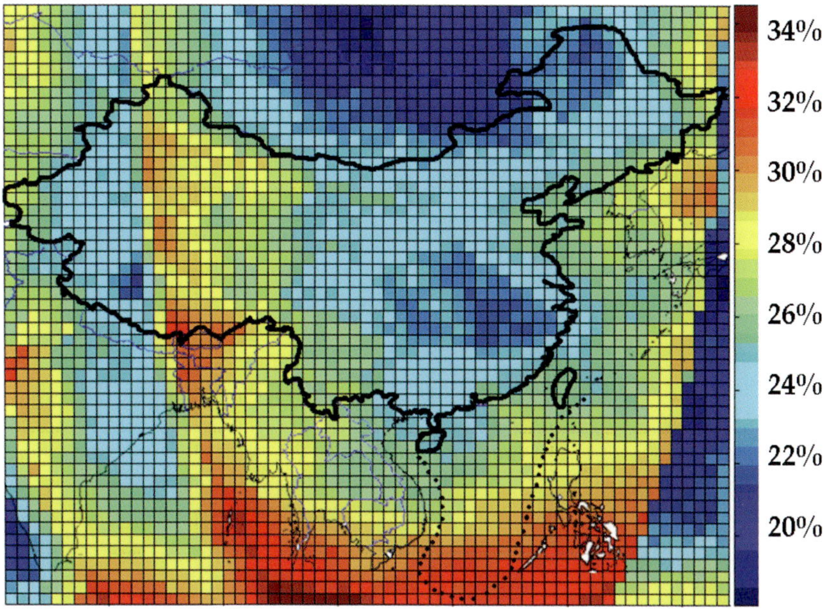

FIGURE 16 Average TDOP reducing in China (compare with GPS).

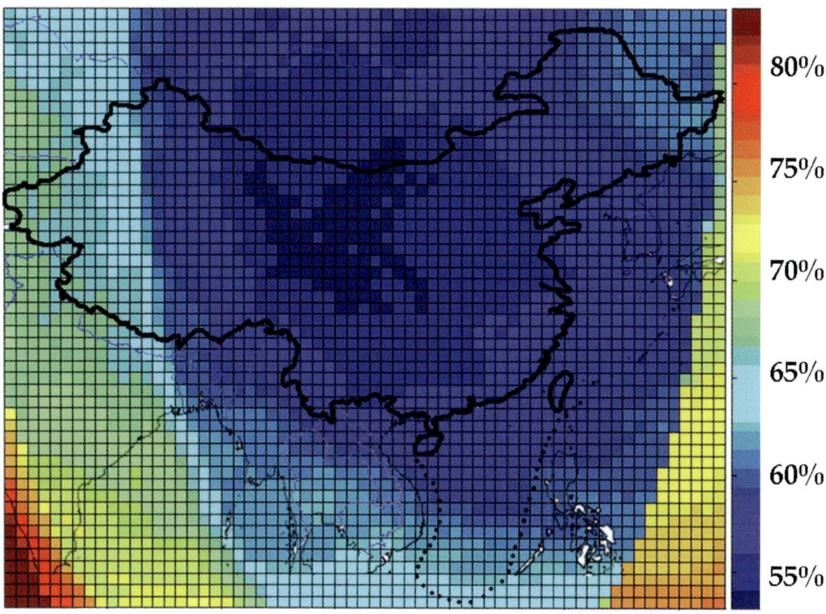

FIGURE 17 Average TDOP reducing in China (compare with BeiDou).

In China, the average PDOP reducing from interoperability constellation to GPS is 29.01 percent; the average TDOP reducing is 25.77 percent; the average PDOP reducing from interoperability constellation to BeiDou is 53.33 percent; and the average TDOP reducing is 61.23 percent. Thus, compared with BeiDou, the interoperability system obtains a 53.33 percent improvement in positioning accuracy and positioning availability, a 61.23 percent improvement in timing accuracy and timing availability; compared to GPS, the interoperability system obtains a 29.01 percent improvement in positioning accuracy and positioning availability and a 25.77 percent improvement in timing accuracy and timing availability.

U.S. Area

Calculate the PDOP and TDOP, shown as Tables 8 and 9.

Calculate PDOP, TDOP of interoperability constellation, GPS constellation, and BeiDou constellation in seven days (a return period) point by point both in the region of United States. Then compare their average to sole system, as shown in Figures 18 and 19.

TABLE 8 PDOP Value of GPS and BeiDou Regional System

	Place	GPS	GPS+BeiDou	Place	GPS	GPS+BeiDou
Mean	Hawaii	1.645	1.391	Los Angeles	1.650	1.536
Max		2.742	2.185		2.265	2.265
Min		1.161	1.051		1.252	1.117
Mean	New York	1.623	1.501			
Max		2.639	2.069			
Min		1.215	1.177			

TABLE 9 TDOP Value of GPS and BeiDou Regional System

	Place	GPS	GPS+BeiDou	Place	GPS	GPS+BeiDou
Mean	Hawaii	0.835	0.690	Los Angeles	0.838	0.773
Max		1.666	1.171		1.365	1.361
Min		0.484	0.457		0.558	0.526
Mean	New York	0.830	0.762			
Max		1.631	1.240			
Min		0.542	0.515			

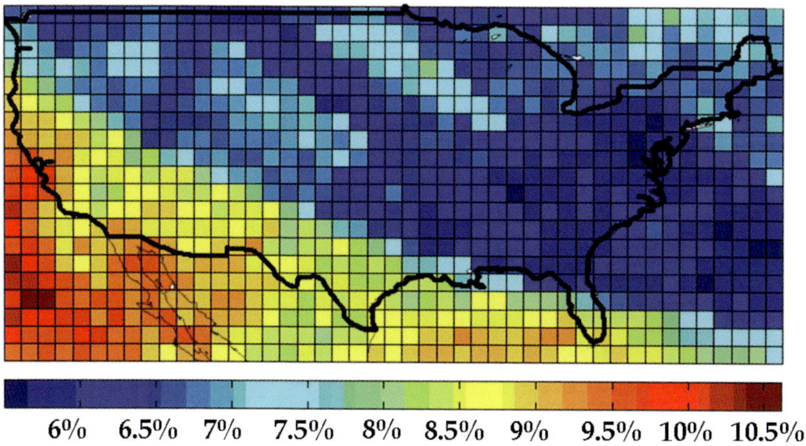

FIGURE 18 Average PDOP reducing in the United States (compare with GPS).

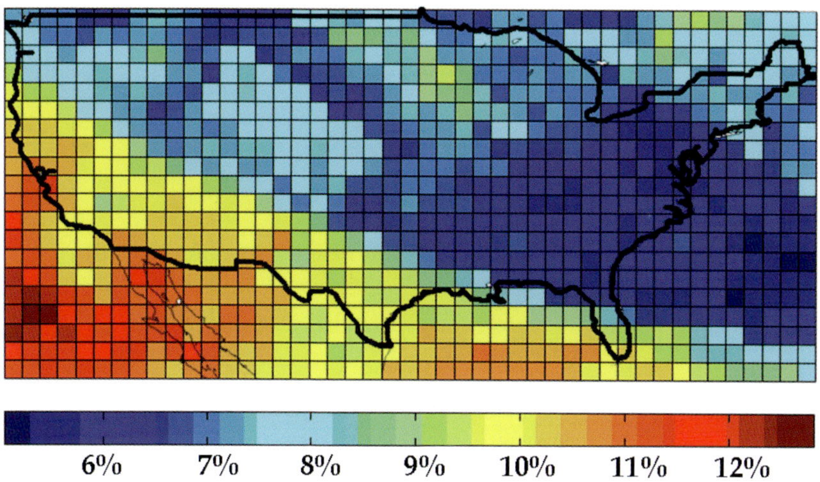

FIGURE 19 Average TDOP reducing in the United States (compare with GPS).

In the United States, the average PDOP reducing from interoperability constellation to GPS is 7.26 percent, and the average TDOP reducing is 8.04 percent. Thus, compared with GPS, the interoperability system obtains a 7.26 percent improvement in positioning accuracy and positioning availability and an 8.04 percent improvement in timing accuracy and timing availability.

Integrity

Service integrity is the probability that the system gives an alarm when there is any malfunction or any error excess threshold value. Service integrity denotes as alarming threshold, alarming period, and hazard misleading information (HMI) probability. Alarming period is the maximum time delay between malfunction beginning and alarm. HMI probability is the probability when the current measure values do exceed the thresholds.

System thresholds of positioning and timing are:

$$\begin{pmatrix} \sigma_P \\ \sigma_t \end{pmatrix} = \begin{pmatrix} PDOP \cdot URE \\ TDOP \cdot \Delta t \end{pmatrix}$$

For the same positioning and timing threshold, decreasing of PDOP and TDOP increases the threshold of URE and Δt. Denote the positioning and timing threshold before interoperability as ε_1, μ_1; which after interoperability are ε_2, μ_2. Because the positioning and timing accuracy are definite, their variance σ_1, σ_2 will be unchangeable. Then:

$$\mathrm{HMI}_1 = P\{(x,y) | x > \varepsilon_1, y > \mu_1\} = \int_{\Omega_1} p(x,y) d\delta \tag{4.12}$$

$$\mathrm{HMI}_2 = P\{(x,y) | x > \varepsilon_2, y > \mu_2\} = \int_{\Omega_2} p(x,y) d\delta \tag{4.13}$$

where $p(x,y)$ is probability density function, for the same $p(x,y)$:

$$\mathrm{HMI}_1 - \mathrm{HMI}_2 = \int_{\Omega_1} p(x,y) d\delta - \int_{\Omega_2} p(x,y) d\delta = \int_{\Omega_1 - \Omega_2} p(x,y) d\delta > 0$$

Thus, the HMI probability before interoperability is smaller than after.

Continuity

Continuity is the ability of avoidance of non-arranged interrupt during the regular operation of a satellite navigation system. Continuity is made up of space signal continuity and positioning service continuity. Space signal continuity is the time ratio of the period of "healthy" signal to a given time interval, which means there is no probability of non-arranged interrupt in a given time interval. Positioning service continuity is the time ratio that the service satisfies the users' positioning accuracy requirement in a given time interval and a given service area. Space signal mean continuity will be calculated by the mean time between failures (MTBF) and mean time to restoration (MTTR) in a non-arranged situation:

$$1-\frac{1}{N}\sum_{i=1}^{N}\frac{\mathrm{MTTR}_i}{\mathrm{MTBF}_i+\mathrm{MTTR}_i}\circ \tag{4.14}$$

where N is the satellite number.

Suppose there are n satellites, when add the kth one, the positioning service continuity will change.

For n+1 satellites, if any satellite is interrupted, the continuity of n+1 satellites is equal to the continuity of n normally working satellites. If no satellite is interrupted, it is the continuity of n+1 normally working satellites.

So, the continuity of n+1 satellites is better than that of n satellites; the continuity of n+2 satellites is better than that of n+1 satellites. Thus, an increased satellite number will enhance positioning continuity.

Brief Summary

Interoperability will promote positioning and timing accuracy; positioning and timing availability; and service integrity and service continuity. Service performance improvement in Chinese region is larger than that in the U.S. region.

Receiver Cost-effectiveness

The structure of a receiver is shown as Figure 20.

Antenna

The frequency difference between B1 and L1 is 14.322 MHz; it is indistinguishable while using omnidirectional antenna both before and after interoperability.

Radio Channel

Some domestic chip factories have already produced the compatible chip of GPS L1 and BeiDou B1. The radio channel, which used to be a troublesome factor, has been greatly solved. According to the market investigation, the civil receiver chip and ADC chip are only 0.7 USD with power consumed 10~30 mW.

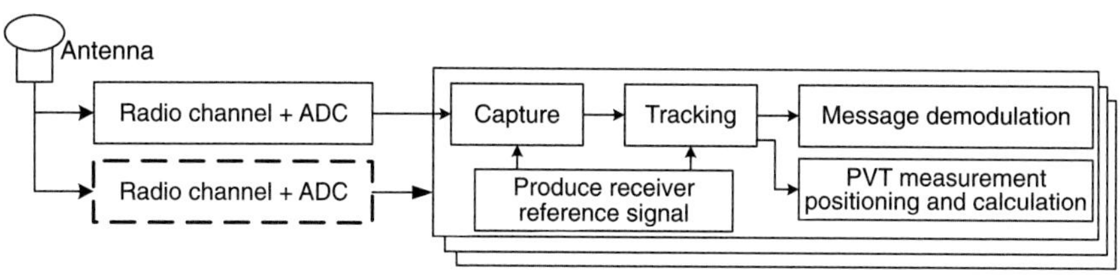

FIGURE 20 Receiver structure.

Capture, Tracking, Demodulation, PVT Measure, and Positioning

The channel of capture, tracking, demodulation, PVT measure, and positioning in the receiver will be added because of the increased numbers of visible satellites; while the cost of each channel is no more than 1 percent of the navigation chip.

FEASIBILITY ANALYSIS OF INTEROPERABILITY BETWEEN BEIDOU B1 BPSK(2) AND GPS TMBOC

Signal Performances Analysis

Correlation Peak

Figure 21 shows the correlation peak curve of BPSK(1), BPSK(2), and TMBOC; where TMBOC has the most sharp correlation peak.

MBOC has a better anti-jamming performance than BPSK(2) and BPSK(1).

Tracking Error

Figure 22 shows that MBOC has a bigger tracking error than BPSK(2) and smaller than BPSK(1). The simulation parameters are listed in Table 4.

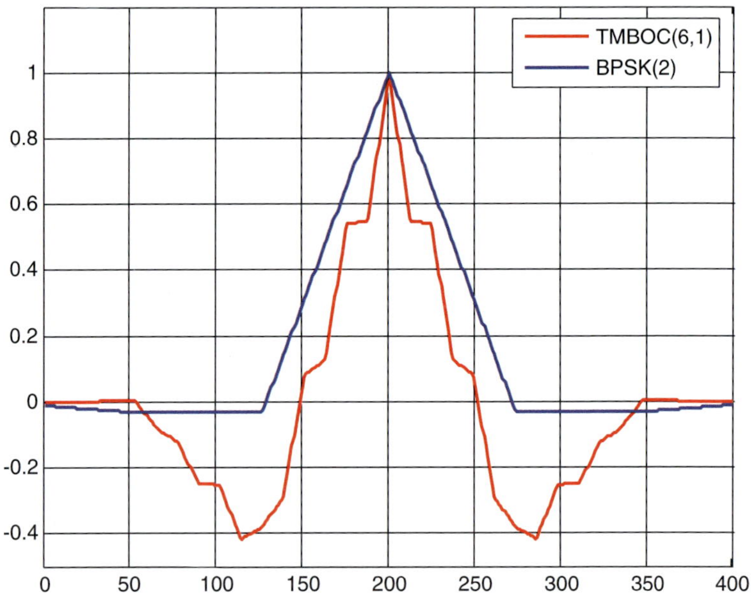

FIGURE 21 Correlation peak of BPSK(1), BPSK(2), and MBOC(6,1,1/1).

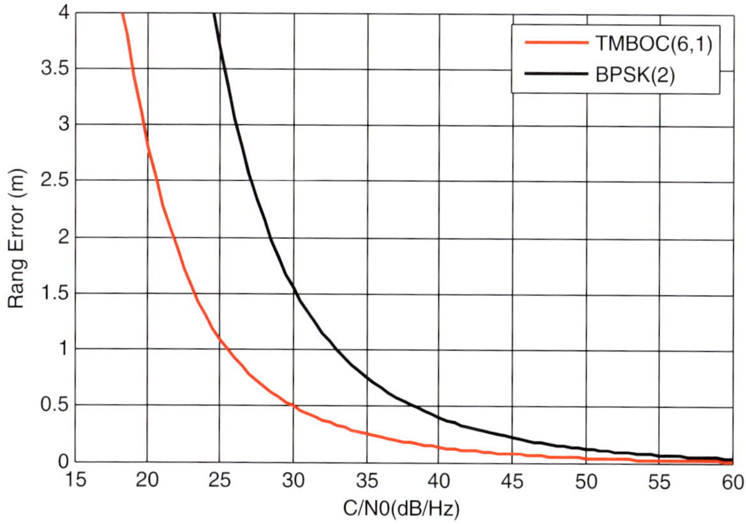

FIGURE 22 Tracking error of BPSK(1), BPSK(2), and MBOC(6,1,1/1).

Multipath

Table 10 and Figures 23 and 24.

Brief Summary

GPS MBOC has better ranging accuracy, anti-jamming, and anti-multipath than BeiDou B1 BPSK(2), but the difference is not large. Interoperability at different frequency improves the anti-jamming and availability of system.

Service Performance

Availability

In China, compared with BeiDou, the interoperability system obtains a 53.33 percent improvement in positioning availability and a 61.23 percent improvement in timing availability; compared with GPS, the interoperability

TABLE 10 Simulation Parameters

B = 30 MHz	a1 = -5 db	d = 1/20
Front receive bandwidth	Ratio of multipath to direct signal	Correlator space

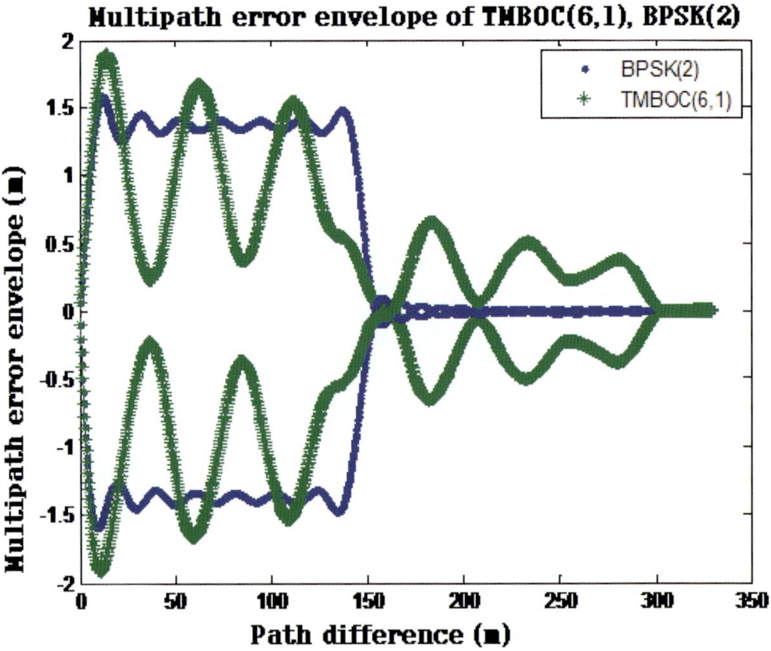

FIGURE 23 Multipath error envelope of BPSK(2) and MBOC(6,1,1/1).

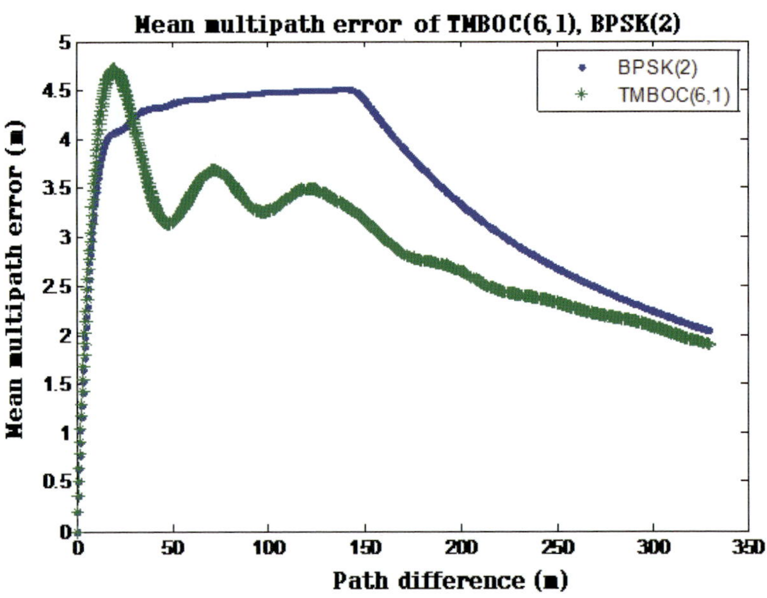

FIGURE 24 Multipath error of BPSK(2) and MBOC(6,1,1/1).

system obtains a 29.01 percent improvement in positioning availability and a 25.77 percent improvement in timing availability.

In the United States, compared with GPS, the interoperability system obtains a 7.26 percent improvement in positioning availability and an 8.04 percent improvement in timing availability.

Integrity

The ranging accuracy of signal in phase two of interoperability is better than in phase one; the integrity detect ability will keep the same level of phase one.

Continuity

Continuity in phase two of interoperability is almost the same as in phase one.

Brief Summary

Comparing the service performance in the second phase of interoperability to that in the first phase, there are improved positioning and timing accuracy, improved positioning and timing availability, enhanced service integrity, and advanced service continuity.

FEASIBILITY ANALYSIS OF INTEROPERABILITY BETWEEN BEIDOU TMBOC AND GPS TMBOC

Signal Performance Analysis

Correlation Peak

The correlation peak curve has a direct relation with ranging accuracy, anti-multipath, and anti-jamming performance. The sharper the peak, the better the performance. Figure 25 shows the correlation peak curves of BPSK(1), BOC(1,1), TMBOC, and CBOC.

Gabor Bandwidth

Figure 26 shows the power spectrum envelopes of TMBOC and CBOC. The RMS bandwidths of different modulations are shown in Table 11.

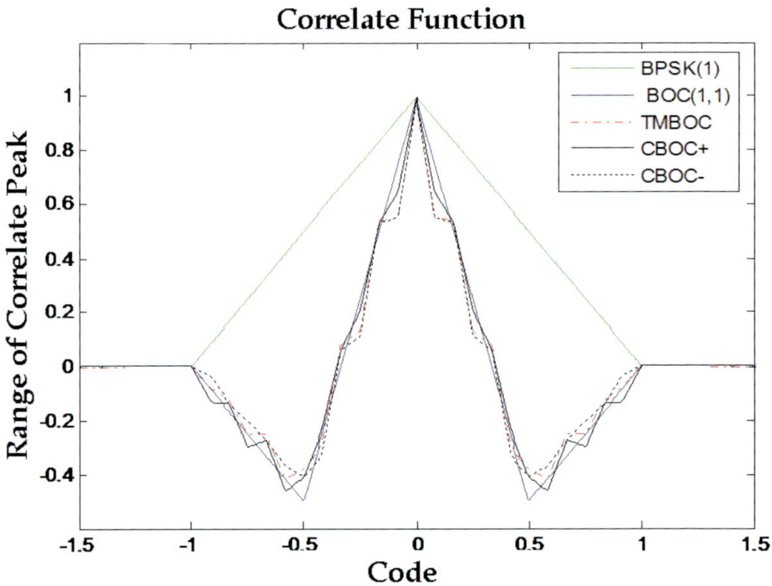

FIGURE 25 Correlation peaks of BPSK(1), BOC(1,1), TMBOC, and CBOC.

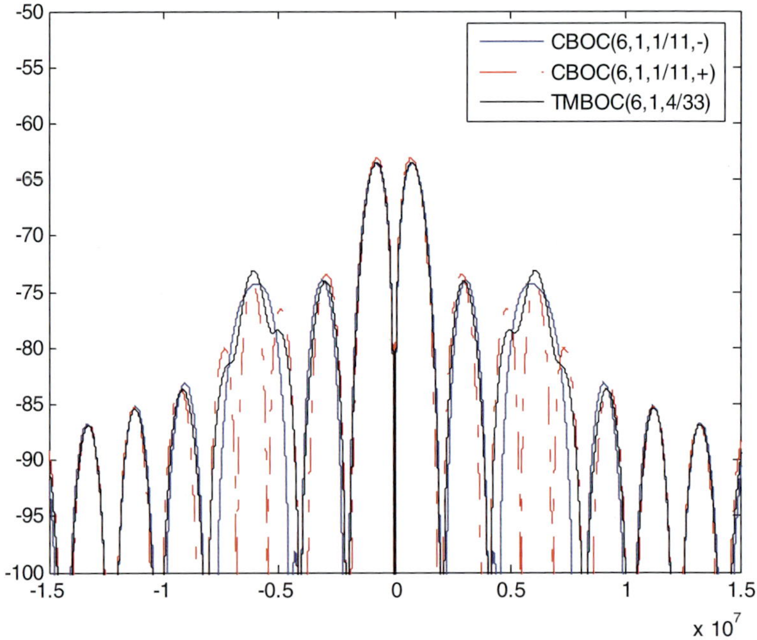

FIGURE 26 Power spectrum envelopes of TMBOC and CBOC.

TABLE 11 Gabor Bandwidth

Modulation	Gabor bandwidth	Modulation	Gabor bandwidth
BPSK(1)	1.4416×10^6	CBOC+(6,1,1/11)	3.0990×10^6
BOCsin(1,1)	2.5194×10^6	CBOC-(6,1,1/11)	3.5348×10^6
BOCcos(1,1)	3.2561×10^6	TMBoc(6,1,1/11)	3.5520×10^6

Tracking Error

Error curves of TMBOC with different correlator spaces are shown in Figure 27 (Table 12).

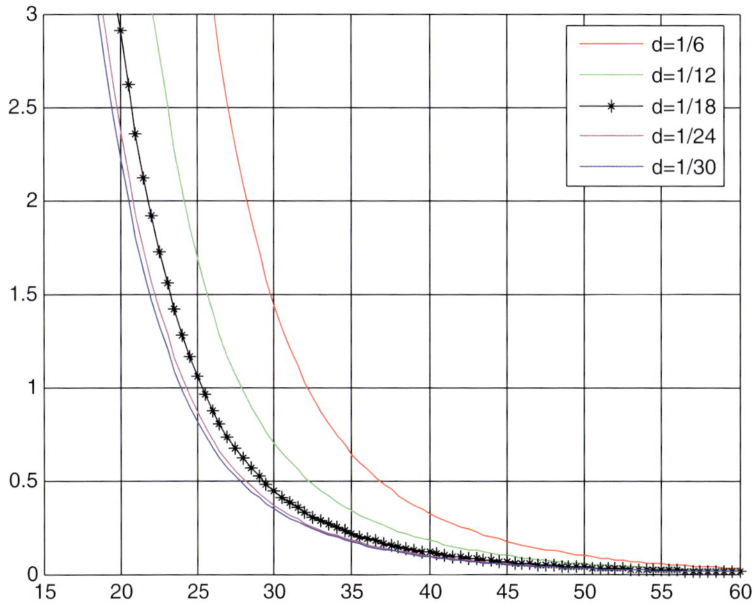

FIGURE 27 TMBOC error curves with different correlator spaces.

TABLE 12 Simulation Parameters

Delt = 5	B = 0.2	IntTime = 0.004
Frequency space of power spectrum	Loop bandwidth	Integral time

Error curves of TMBOC with different receiver front bandwidths are shown in Figure 28 (Table 13)

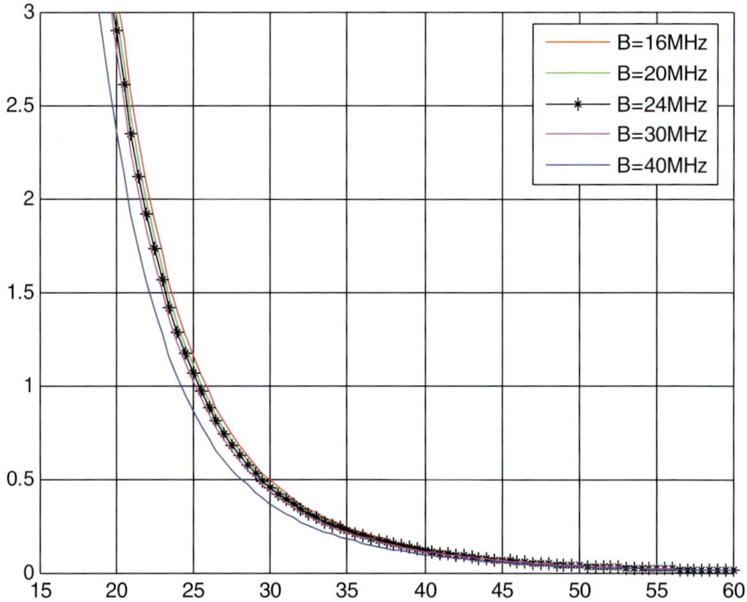

FIGURE 28 TMBOC error curves with different receiver front bandwidths.

TABLE 13 Simulation Parameters

Delt = 5	B = 0.2	IntTime = 0.004
Frequency space of power spectrum	Loop bandwidth	Integral time

Error curves of TMBOC with different receiver loop bandwidths are shown in Figure 29 (Table 14).

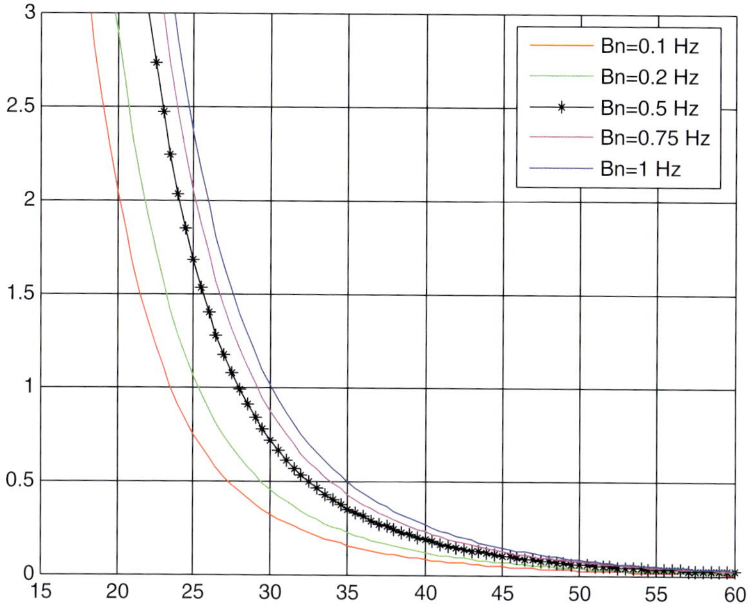

FIGURE 29 TMBOC error curves with different receiver loop bandwidths.

TABLE 14 Simulation Parameters

Delt = 5	IntTime = 0.004	Dllw = [0.1 0.2 0.5 0.75 1]	d = 1/24
Frequency space of power spectrum	Integral time	Loop bandwidth	Correlator space

Error curves of TMBOC with different integral time are shown in Figure 30 (Table 15).

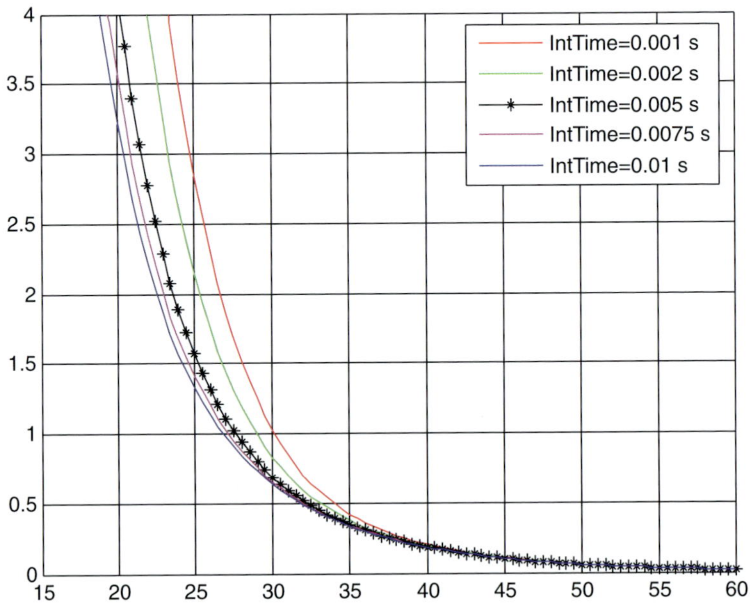

FIGURE 30 TMBOC error curves with different integral time.

TABLE 15 Simulation Parameters

Delt = 5	Dllw = 0.5	Dllw = [0.1 0.2 0.5 0.75 1]	d = 1/24
Frequency space of power spectrum	Loop bandwidth	Loop bandwidth	Correlator space

Error curves of signals with different modulations are shown as Figure 31 (Table 16).

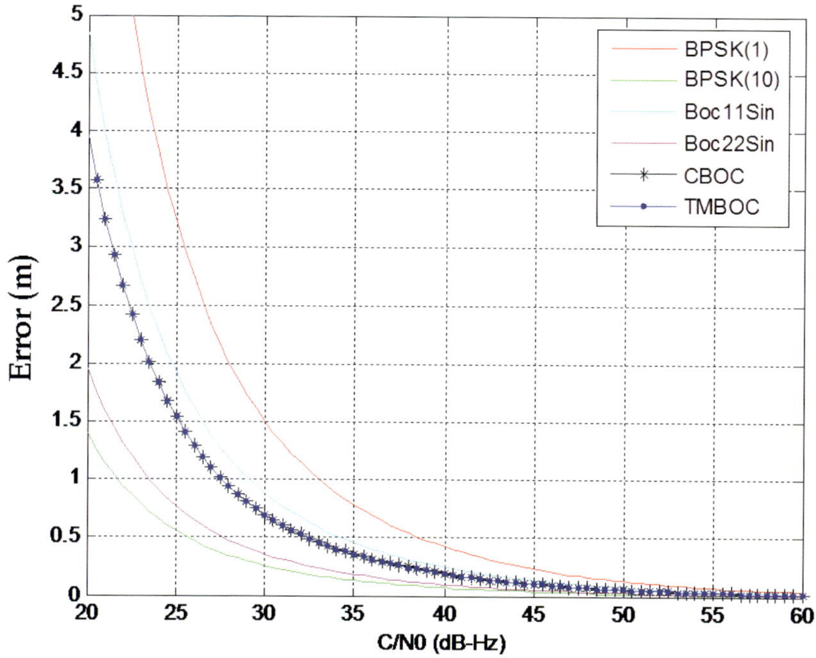

FIGURE 31 Error curves of signals with different modulations.

TABLE 16 Simulation Parameters

Delt = 5	Dllw = 0.5	IntTime = 0.004 s	d = 1/24
Frequency space of power spectrum	Loop bandwidth	Integral time	Correlator space

Multipath

Generally, we can get different multipath performance for each kind of navigation signal based on its multipath error and multipath running average error.

Assume that receiver front bandwidth is 30 MHz, with correlator space of 1/20 chip and the ratio of multipath to direct path of -6dB. Figure 32 shows the envelop curves of multipath average errors for BPSK(10), BOC(1,1), BOC(2,2), TMBOC Pilot, and CBOC Pilot.

If we put the modulated signal with best performance in the first place, and that with worst performance in the end, then we can easily get the following results: BPSK(10), BOC(2,2), TMBOC(6,1,4/33), CBOC-(6,1,1/11), BOC(1,1), BPSK(1); where the difference between TMBOC(6,1,4/33) and CBOC-(6,1,1/11) is very small.

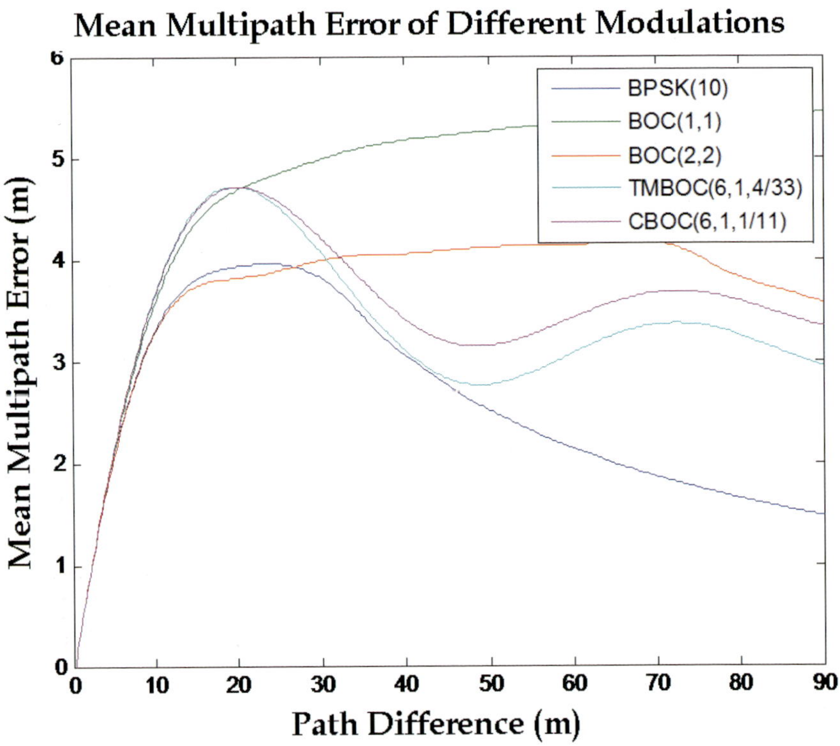

FIGURE 32 Multipath running average error.

Brief Summary

Both BeiDou B1 signal in phase 3 and GPS modernized signals adopt the design of TMBOC(6,1,4/33) + BOC(1,1). The particular technologies of time division in sub-carrier wave, second coding for ranging codes, channel separation, and message coding make the TMBOC signal have a better acquisition performance, better tracking performance, better demodulation performance, better anti-jamming performance, and better anti-multipath performance.

Service Performance

In phase 3 of interoperability between BeiDou and GPS, not only did the signals of BeiDou system change, but its satellite constellation also changed (Figures 33–38).

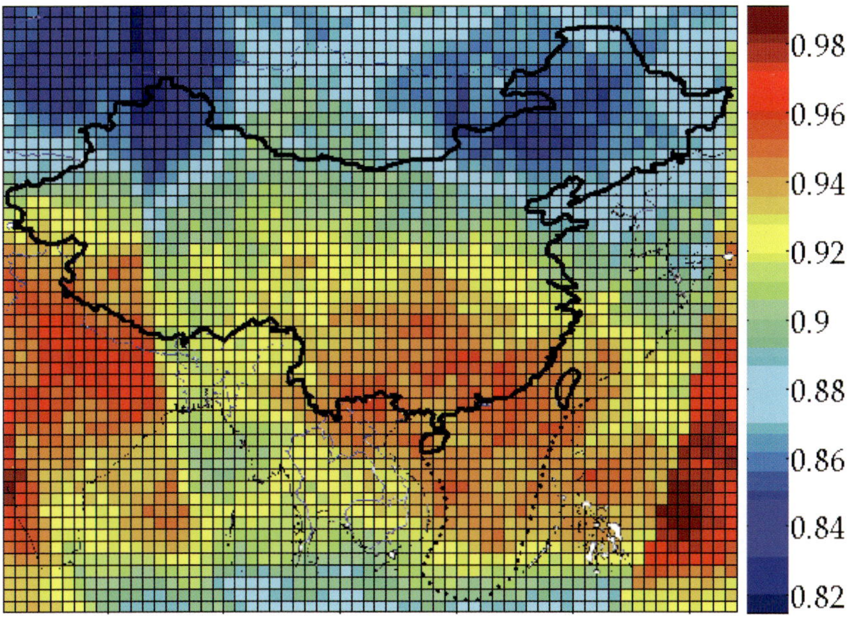

FIGURE 33 PDOP in China.

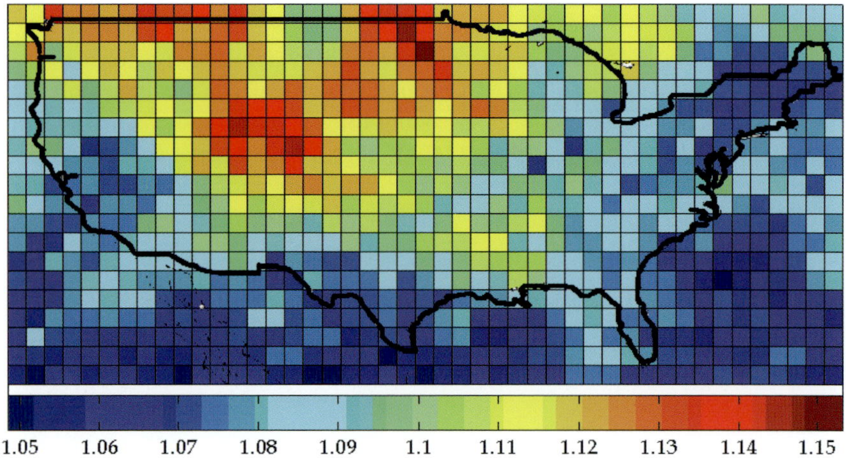

FIGURE 34 PDOP in the United States.

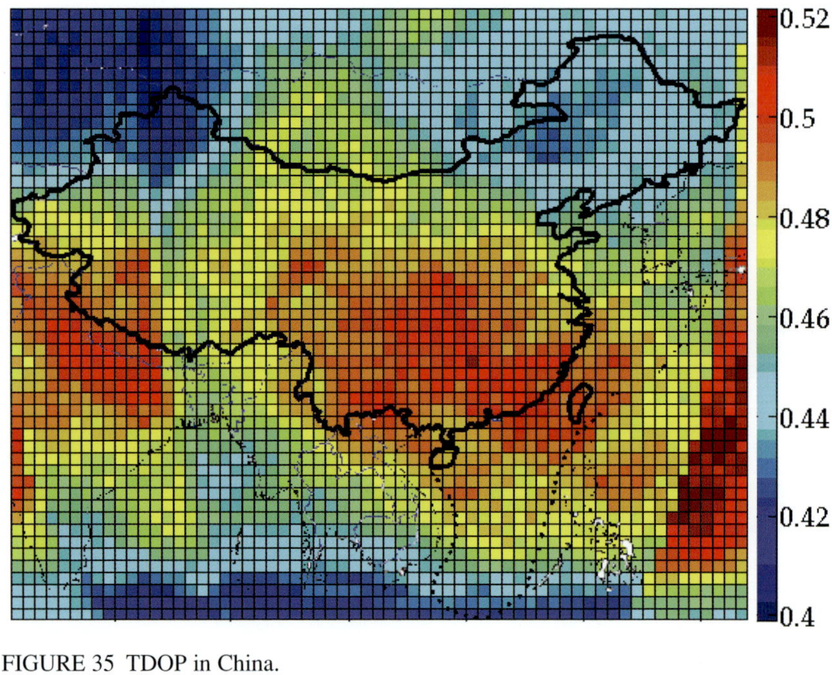

FIGURE 35 TDOP in China.

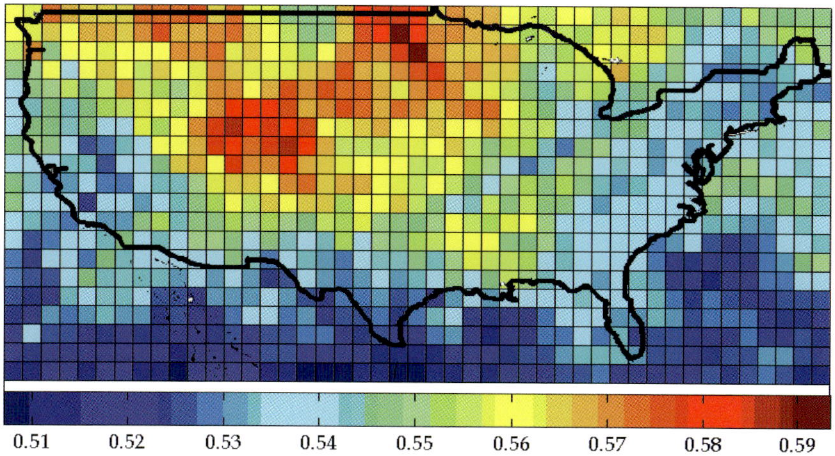

FIGURE 36 TDOP in the United States.

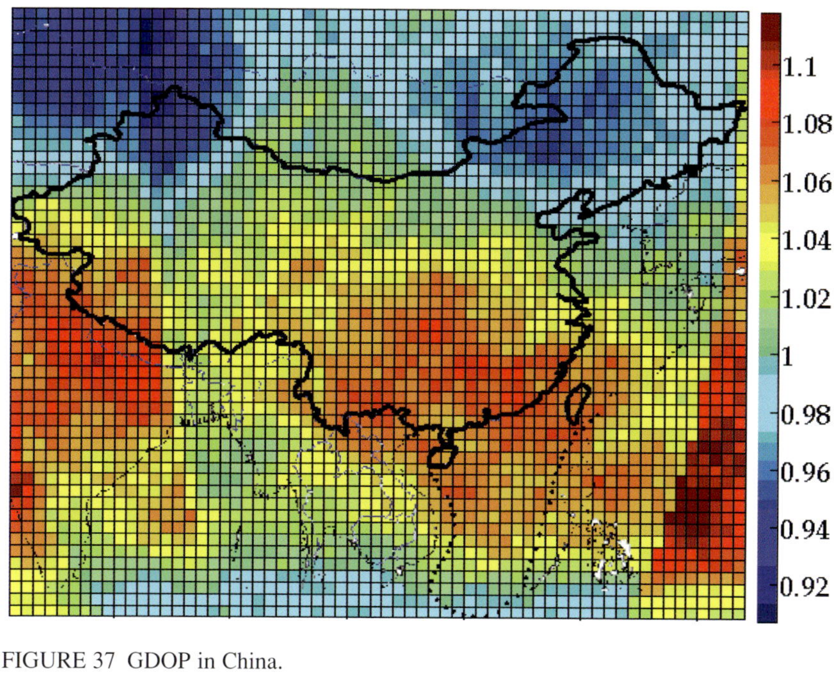

FIGURE 37 GDOP in China.

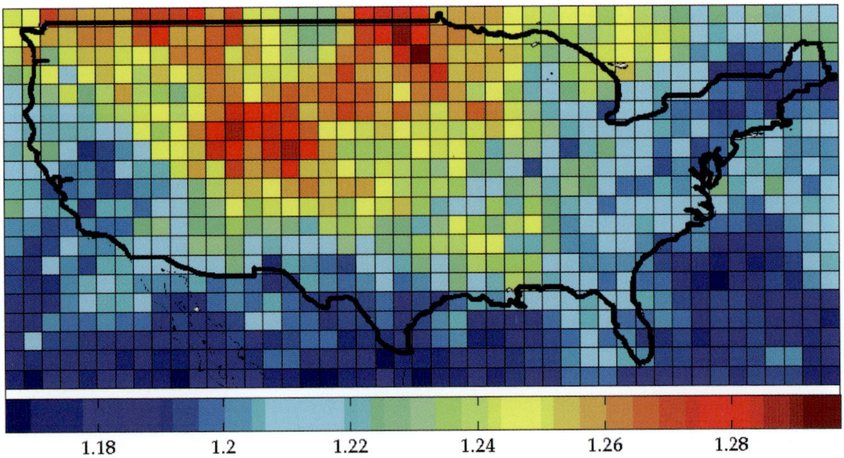

FIGURE 38 GDOP in the United States.

Accuracy

In China, for an interoperability constellation of BeiDou + GPS, the average PDOP reduction is 30.14 percent compared with that of GPS, while the average TDOP reduction is 28.76 percent. When it is compared with BeiDou, the average PDOP reduction is 17.75 percent, and the average TDOP reduction is 20.14 percent. Thus, compared with GPS, the interoperability system obtains a 30.14 percent improvement in positioning accuracy and a 28.76 percent improvement in timing accuracy. When it is compared with BeiDou, the interoperability system obtains a 17.75 percent improvement in positioning accuracy and a 20.14 percent improvement in timing accuracy.

In the United States, for an interoperability constellation of BeiDou + GPS, the average PDOP reduction is 35.10 percent compared with that of GPS and the average TDOP reduction is 36.66 percent. When it is compared with BeiDou, the average PDOP reduction is 35.86 percent and the average TDOP reduction is 37.97 percent. Thus, compared with GPS, the interoperability system obtains a 35.10 percent improvement in positioning accuracy and a 36.66 percent improvement in timing accuracy. When it is compared with BeiDou, the interoperability system obtains a 35.86 percent improvement in positioning accuracy and a 37.97 percent improvement in timing accuracy.

Availability

In China, for an interoperability constellation of BeiDou + GPS, the average PDOP reduction is 30.14 percent compared with that of GPS and the average TDOP reduction is 28.76 percent. When it is compared with BeiDou, the average PDOP reduction is 17.75 percent and the average TDOP reduction is 20.14 percent. Thus, compared with GPS, the interoperability system obtains a 30.14 percent improvement in positioning availability and a 28.76 percent improvement in timing availability. When it is compared with BeiDou, the interoperability system obtains a 17.75 percent improvement in positioning availability and a 20.14 percent improvement in timing availability.

In the United States, for an interoperability constellation of BeiDou and GPS, the average PDOP reduction is 35.10 percent and the average TDOP reduction is 36.66 percent. When it is compared with BeiDou, the average PDOP reduction is 35.86 percent and the average TDOP reduction is 37.97 percent. Thus, compared with GPS, the interoperability system obtains a 35.10 percent improvement in positioning availability and 36.66 percent improvement in timing availability. When it is compared with BeiDou, the interoperability system obtains a 35.86 percent improvement in positioning availability and a 37.97 percent improvement in timing availability.

Integrity

The ranging accuracy of interoperable signal in phase 3 is better than that of phase 2; while the GDOP in phase 3 is better than that of phase 2. Thus, the integrity detection ability will be improved.

Continuity

Increase of the number of BeiDou satellites also adds the number of interoperability constellation satellites in phase 3, which will lead to a better continuity than that of phase 2.

Brief Summary

In phase 3, there will be a smaller ranging error, advanced URE, more visible satellites, and lower positioning and timing threshold. As a result, service availability, integrity, and continuity will be improved. So we can draw the conclusion that, if BeiDou and GPS could achieve interoperability, then there will be a great improvement in the service performance of both systems.

SUMMARY

From the analysis above, we can see that:

1. BeiDou and GPS could achieve interoperability.
2. Interoperability between BeiDou and GPS will be of great benefit to the improvement in service performance.

3. With the development of each system, interoperability between the two systems will deepen.
4. Interoperability could be achieved between BeiDou B2 signal and GPS L5 signal as well, so dual-frequency interoperability can also be realized.

The Interchangeability Problem: Signals, Coordinate Frames, and Time

RITA M. LOLLOCK and THOMAS D. POWELL
The Aerospace Corporation

THOMAS A. STANSELL
Stansell Consulting

INTRODUCTION

The Global Positioning System (GPS) and the GLObal NAvigation Satellite System (GLONASS) were developed independently. Although similar in certain respects (e.g., constellation design and dual L-band frequencies), there was no intent by the developers to make the systems interoperable. The first known receiver to use both GPS and GLONASS signals for navigation was a prototype developed by Magnavox and delivered to Lincoln Laboratory in 1990. To achieve satisfactory navigation results the receiver had to compensate for differences in system time, coordinate frame, signal frequencies, spreading codes, message formats, and separate filter delays within the receiver. Nevertheless, this and subsequent commercial receivers were able to demonstrate that two rather dissimilar systems could be used together to achieve better results than with either system alone, especially in difficult reception environments, even though the systems were far from "interchangeable."

In recent years new systems have been proposed and are being developed, including Compass, Galileo, and QZSS (Quasi-Zenith Satellite System). Also, both GPS and GLONASS are being modernized with new signals. These systems are now referred to collectively as Global Navigation Satellite Systems, or GNSS. Fortunately, the international climate has changed substantially such that "compatibility" and "interoperability" are important topics at multilateral meetings such as the International Telecommunication Union (ITU) and the International Committee on GNSS (ICG) as well as at bilateral meetings between system providers. Compatibility between systems is vital to ensure that signals from one system do not unacceptably degrade the performance of other systems. Primarily this means that compatible systems must provide signals with similar maximum and minimum received power levels.

The internationally accepted definition of interoperability is the "ability of open global and regional satellite navigation and timing services to be used together to provide better capabilities at the user level than would be achieved by relying solely on one service or signal." As a result of international cooperation, new and modernized GNSS signals will have characteristics that substantially improve interoperability as compared with the original GPS and GLONASS signals.

Some have suggested that the goal should be not only interoperability but also interchangeability, meaning that there should be no discernible differences between the signals from different systems. This paper addresses the problems of achieving full interchangeability of signals while also showing why the remaining differences will not affect users.

IDEAL INTERCHANGEABILITY

It is instructive to realize that the current GPS constellation is populated with four different types of satellites with four significantly different designs from different manufacturers, as shown in Figure 1. These are the GPS Block IIA satellites built by Rockwell International (now Boeing), the GPS Block IIR and the Block IIR-M satellites built by Lockheed Martin, and the GPS Block IIF satellites built by Boeing. Later in this decade a fifth type of satellite, the GPS Block III being designed by

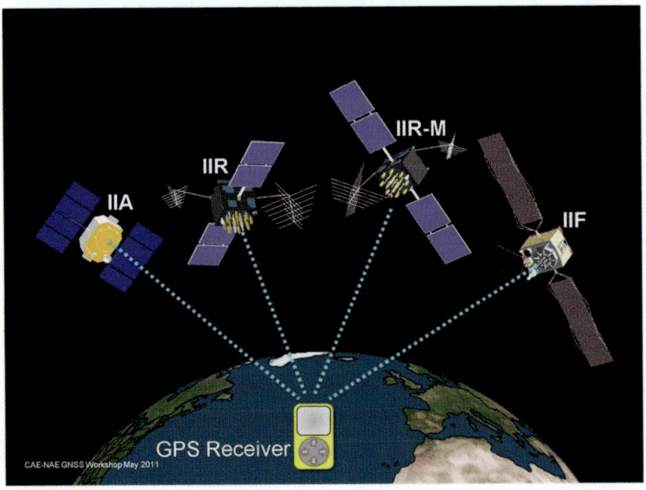

FIGURE 1 GPS receivers track four different satellite types.

Lockheed Martin, will join the constellation. The Block IIR-M added three new signals, the Block IIF added a fourth, and the Block III will add a fifth.

From a current user's perspective, signals from different types of GPS satellites are completely interchangeable. This is because "backward compatibility" is required for each new type of satellite. Each legacy signal shares a common signal structure and the same orbit message parameters, geodesy, tracking stations, and GPS time. They are part of a unified system.

This type of interchangeability is not practical for other global navigation satellite systems. Each of these systems is separately owned, developed, and operated. They employ different signal structures, message formats, orbit characteristics, geodesy, tracking networks, control system software, and system time. Therefore, different GNSS signals are not inherently interchangeable. This paper evaluates whether current international cooperation on interoperability will produce true interchangeability from the user perspective. In other words, will users experience differences between systems?

INTEROPERABILITY PROGRESS AND STATUS

As a result of international bilateral and multilateral meetings, the advantages of GNSS interoperability have been recognized and major strides made toward achieving the goal. A particularly important achievement is that right hand circular polarization, code division multiple access (CDMA) modulation, and two identical GNSS carrier frequencies—1176.45 MHz and 1575.42 MHz—have been adopted by GPS, Compass, Galileo, and QZSS. Figure 2 illustrates each GNSS

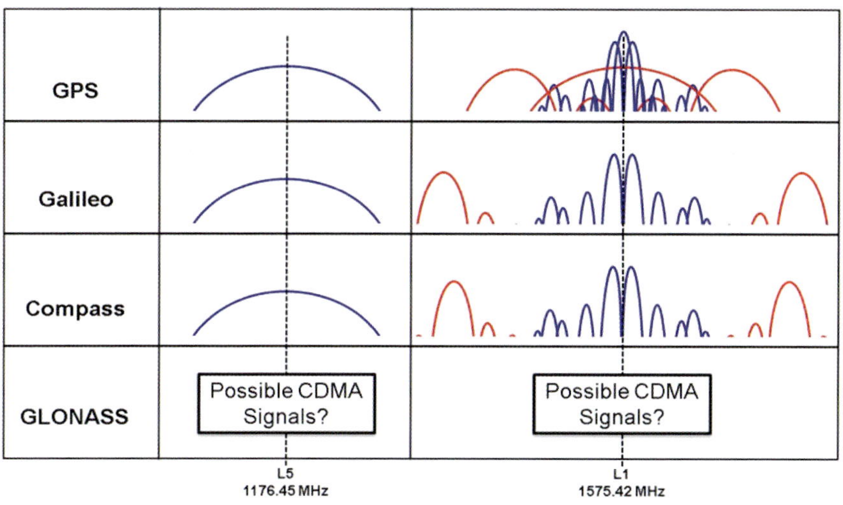

FIGURE 2 GNSS L1 and L5 signal plans.

provider's signals centered on these two frequencies. GLONASS is considering these possibilities as well as other options that could provide the advantage of frequency diversity.

Differences in signal structure do exist, however. Some differences are because each constellation of satellites is different, and messages describing the orbits must be appropriate for each system. Beyond that, for example, the data rate on both the GPS L1C and L5 signals is 50 bits per second (bps) whereas the Galileo E1 OS data rate is 125 bps and the E5a data rate is 25 bps. GPS L1C employs a time multiplexed binary offset carrier (BOC) or TMBOC waveform, whereas the Galileo E1 OS employs a composite BOC (CBOC) waveform. The length of the GPS L1C spreading code is 10 ms or 10,230 chips, whereas the Galileo E1 OS spreading code length is 4 ms or 4,092 chips. Each signal has been developed to serve different objectives. In spite of these differences, interoperability is enhanced because of identical center frequencies and spectral characteristics.

However, users are not likely to notice any of these differences regardless of which combination of signals is being used at the moment. This is because user equipment is being designed to combine all types of signals to give one position, velocity, and time (PVT) solution. If different signals are known to have different levels of accuracy, they will be appropriately weighted in the combined solution. Users will not know or care about the "nationality" of the signals, but they will enjoy the benefits of improved availability and accuracy due to the larger number of signal sources, as shown by the "Multi-GNSS" receiver in Figure 3.

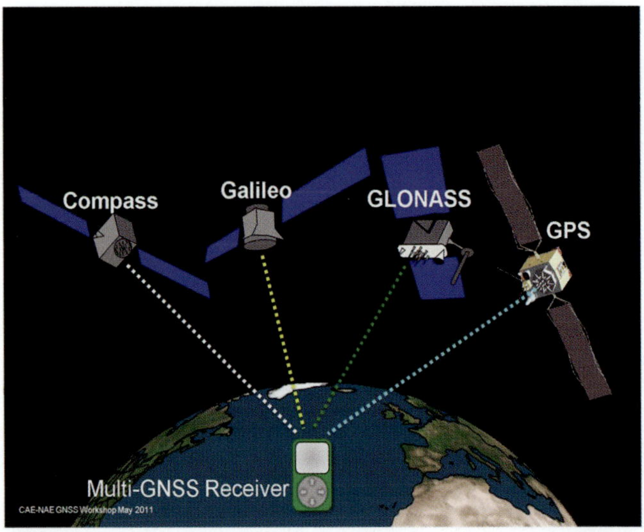

FIGURE 3 Transparent multi-GNSS interoperability is achievable through system and receiver design.

The key to understanding how user equipment will compensate for the signal differences is to recognize two classes of difference, i.e., analog and digital. Because of Moore's law, the cost of digital electronics has continued to plummet while performance has greatly increased over previous decades, and the trend is expected to continue. On the other hand, dimensions of analog components such as antennas and filters are constrained by the fixed signal wavelength and have not experienced as much improvement. Therefore, it is important that signal characteristics that affect analog components be as common as possible. These include antenna polarization, center frequency, and waveform or spectrum. Fortunately, the standards for GPS, Compass, Galileo, and QZSS are right-hand circular polarization, 1575.42 MHz and 1176.45 MHz center frequencies, and a common signal spectrum at each center frequency. This allows use of common analog receiver components for each of these signals.

All other signal differences, such as the spreading codes, data rates, message structures, etc., are handled by digital electronics and software. It is not that these differences are unimportant, because receiver designers must accommodate each one, but the differences are invisible to the user because they are hidden within a special-purpose digital chip and internal software. Because digital chips and memory have become so cheap and so competent, a receiver using signals from all GNSS providers will cost no more than a receiver using signals from only one. In fact, because competition will drive every manufacturer to offer the best possible performance and features, it is likely that Multi-GNSS receivers will become the norm in the future.

SYSTEM TIME AND GEODETIC COORDINATE SYSTEMS

Concern has been expressed about how to achieve a common time standard and common spatial coordinate systems. There are two very effective approaches to resolving these issues.

First, time differences can be resolved by software in individual receivers. This was demonstrated in 1990 by the first GPS/GLONASS navigation receiver. System time differences are calibrated very precisely by including a time offset parameter in the navigation solution matrix. In other words, instead of solving for three position coordinates plus a receiver time offset by observing four or more signals from one system, it is straightforward to additionally solve for a system time offset by observing five or more signals from two systems. Some think this effectively removes one satellite from the observation matrix. However, that is an oversimplification. Because the time difference between two systems changes very slowly, the fifth variable can be determined quickly and then highly filtered. Once established and change is constrained, every satellite signal is able to participate in the navigation solution. The system time difference calibration can also be remembered and used immediately when a receiver is next switched on.

The difference between geodetic coordinate frames can be established very precisely by a network of receivers tracking signals from each system. The International GNSS Service (IGS) is an excellent example of this capability. Therefore, the offset between geodetic coordinate systems can be determined precisely and provided to user equipment in many ways. Some GNSS may choose to include coordinate offsets in their broadcast messages. These messages also will be provided over the Internet, transmitted by cell phone networks, or included as part of receiver software updates.

The other approach is for GNSS providers to operate with common time and geodetic standards. This is easier to achieve with the geodetic reference systems, because all providers consider the International Terrestrial Reference System (ITRS), which is established and updated by the IGS, to be the best available Earth representation. Therefore, every provider has moved their internal reference coordinate system ever closer to ITRS. As a result, most systems already agree or soon will agree with ITRS to within a few centimeters. Not only is this negligible for the vast majority of users, but also high-precision users employ differential corrections that eliminate the effect of any residual difference. Therefore, except for scientific geodesists, we can consider this problem solved.

System time differences are harder to resolve but easy to observe. Just as each user equipment can observe, solve for, and eliminate the effect of system time differences, networks can do the same. This ability would be enhanced by the interoperability agreements. With identical center frequencies and signal spectra, reference stations can precisely measure the time offset between satellites of different systems. Although providers could use this information to drive their own system clocks to agree with any one of the other systems, it would be preferable to establish an international GNSS time standard against which each system could measure its time and drive any difference toward zero.

Measured time differences can be communicated to user equipment in many ways. They could be included as part of the satellite message or provided by the Internet or over cell phone communications. Provisions already have been made for some systems to provide time offsets relative to one or more other systems. Multiple systems have announced initial plans to use GPS Time for this purpose. Knowing the offset in advance will help users in locations with limited visibility to obtain a position fix quickly with a few signals from several different systems. However, it should be noted that once many signals can be tracked simultaneously, the user equipment will measure the time differences more precisely on its own.

SPECIFICATIONS AND STANDARDS

As stated above, manufacturers already are designing, if not shipping, receivers capable of combining signals from many GNSS. However, including a system is not possible until its signal characteristics are well understood. This requires early publication by each GNSS provider of accurate signal specifications. Manufacturers will guess at the specifications and provide for future software updates, but only with accurate signal specifications will use of a particular system be assured. There also should be a way for manufacturers to ask questions and receive prompt replies from the system providers.

System performance standards help manufacturers anticipate how best to combine signals. Ultimately, of course, signal combination algorithms will be optimized by practical measurements.

When signals are used for safety of life applications, it is particularly important for providers not only to define the signals and the expected performance standards but also to rapidly issue service notifications as is done by GPS Notice Advisories to Navstar Users (NANUs). For example, the International Civil Aviation Organization (ICAO) requires specific service commitments and ongoing support before a GNSS can be accepted for use.

SUMMARY AND CONCLUSIONS

GPS was declared fully operational on April 27, 1995. More than 16 years later its performance continues to improve and the number of applications and users continues to expand exponentially. As phenomenal as this has been, GNSS is on the verge of revolutionary improvements as GPS signals complete their modernization, the GLONASS constellation completes its replenishment, and new systems such as Compass, Galileo, and QZSS become operational. In the future, when three GNSS frequencies become generally available, wide-area 10 cm phase-based navigation will proliferate.

Availability and accuracy for all forms of navigation will improve dramatically because the number of visible satellites will more than double. Figure 4 illustrates this for an environment with significant visibility obstructions, such as an urban canyon. Using only the five GPS satellites in view results in a position dilution of precision (PDOP) of 4.84. Adding the four Compass satellites in view improves the PDOP to 3.92. Adding the two Galileo satellites gives a PDOP of 2.37. With three GLONASS satellites, the PDOP becomes 2.16. The clear message of Figure 4 is that having many interoperable satellites will provide PNT availability even in difficult environments and with much better accuracy than possible with only one system.

International GNSS cooperation has improved dramatically so that users will experience the full benefits of interoperable signals. Because receivers will seamlessly combine signals with different digital characteristics, the user perception will be that all GNSS signals are fully interchangeable.

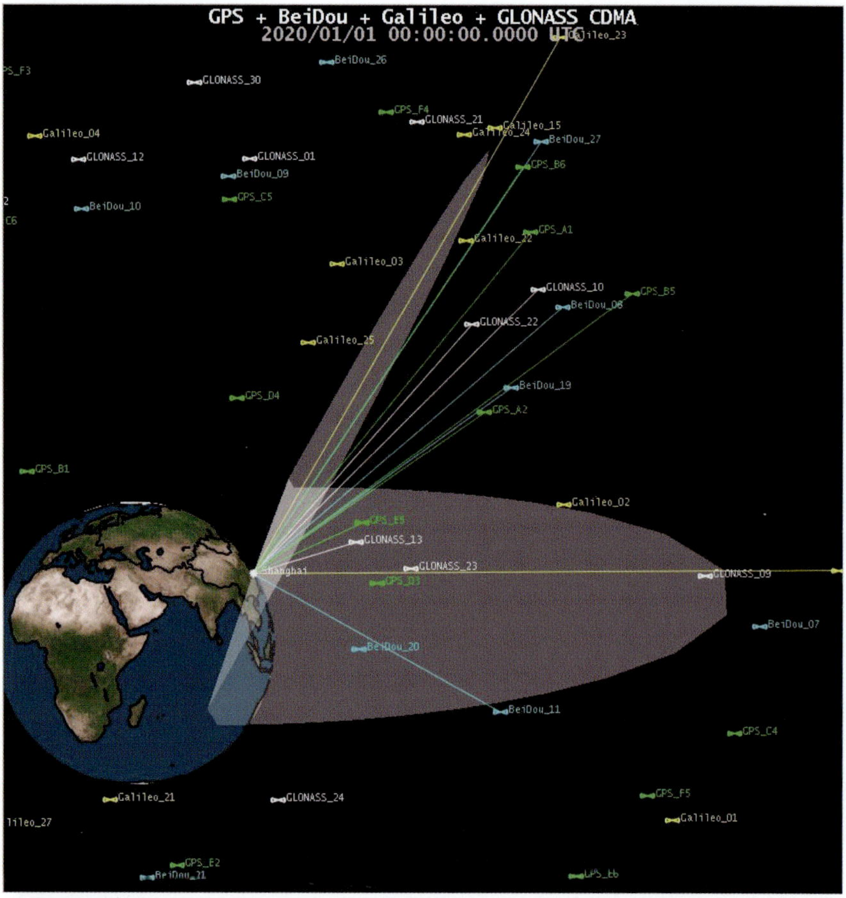

FIGURE 4 Better user accuracy is achievable via better position dilution of precision (PDOP) due to multi-GNSS interoperability.

COMPASS/BeiDou Coordinate and Time Reference Systems

YANG YUANXI and TANG JING
China National Administration of GNSS and Applications (CNAGA)

HAN CHUNHAO
Beijing Global Information Center of Application and Exploration

ABSTRACT

Satellite navigation needs coordinate and time references. It is impossible to realize interoperability for multiple GNSS systems without a consistent reference coordinate system and time system. Compass/BeiDou satellite navigation system follows the rules of compatibility and interoperability defined by the International Committee of GNSS (ICG). The coordinate system of BeiDou is aligned to the China Geodetic Coordinate System 2000 (CGCS 2000), which is aligned to ITRS. BeiDou system time (BDT) is an internal, continuous navigation time scale, without leap second. BDT is linked to the national UTC(k), which is consistent to UTC. The status and existing problems and future developments are described.

BEIDOU COORDINATE SYSTEM (BDC)

GNSS provides position and time service, while it needs reference coordinate and time. Multi-GNSS will do nothing without the same datum of coordinate and time system. In theory, all of the GNSS coordinate reference system should be referred to the same geocentric reference coordinate system, which is beneficial to the satellite navigation providers, the companies/producers of the GNSS receivers, and the users. However, the same reference coordinate system does not mean the same reference frame, which is unnecessary and impossible for system providers at the present stage for several reasons. First, different nations or different continents may realize their own reference coordinate system by using different coordinate frames. Second, different coordinate frames have different uncertainty and different systematic errors. Third, the offsets between different coordinate frames may be detected by some common stations with different satellite positioning coordinates. Fourth, the satellite orbit parameters may reflect the reference coordinate, which can be monitored by multi-GNSS receivers. Fifth, the reference coordinate system should include the dynamic deformation information, and the update of the reference frame should have standards, otherwise additional satellite orbit errors may be introduced in the satellite navigation systems.

Foundation of BDC

BDC is consistent with CGCS 2000, which in turn is consistent with the International Terrestrial Reference System (ITRS), and is a global, three-dimensional, right-handed, orthogonal, geocentric reference system. CGCS is realized by China Terrestrial Reference Frame 2000 (CTRF2000), which is referred to ITRF97 with the epoch of 2000.0 (Yang, 2009). The main parameters of the reference ellipsoid are nearly the same as those defined by ITRS with the semi major axis (a), second degree harmonic coefficient (J2), and the mean angular velocity of the Earth (ω) the same as those of the ellipsoid of GRS1980. The gravitational mass (GM) constant adopts the value of WGS-84 (Wei, 2003).

It should be pointed out that there exists little difference between the BDC coordinate system and the GPS geodetic system (GGS-84). The flattening of the ellipsoid of BDC is f = 1/298.257222101, which is different from that of GPS with f = 1/298.257223563. In theory, the biggest offset introduced by the two flattening ratios is within 1 millimeter in the equator. This kind of difference may be neglected, that is, the coordinate references of BeiDou system and GPS are consistent.

BDC is realized by BeiDou Terrestrial Reference Frame (BTRF), which is connected to CTRF 2000. The CTRF 2000 adopts the same standards as those of the ITRF. There are 47 International GNSS Service (IGS) stations evenly distributed over the world (Figure 1), including those IGS stations in China: BJFS, CHAN, KUNM, LHAS, SHAO, URUM, WUHN, and XIAN (Figure 2) that were used in the integrated adjustment of CTRF. Consistency in definition and diversity in realization will be beneficial for compensating the systematic errors.

Three levels of CTRF are included: the first level is the IGS and continuously operating reference stations (CORS) (28 stations) at the 3 mm level; the second level is the 2000 National GPS Network with about 2,500 stations at 3 cm level (Yang et al., 2007, 2009); and the third level is the combined adjusted network of astro-geodetic network and 2000 National GPS Network with 50,000 stations at 3 dm level (Yang et al., 2005).

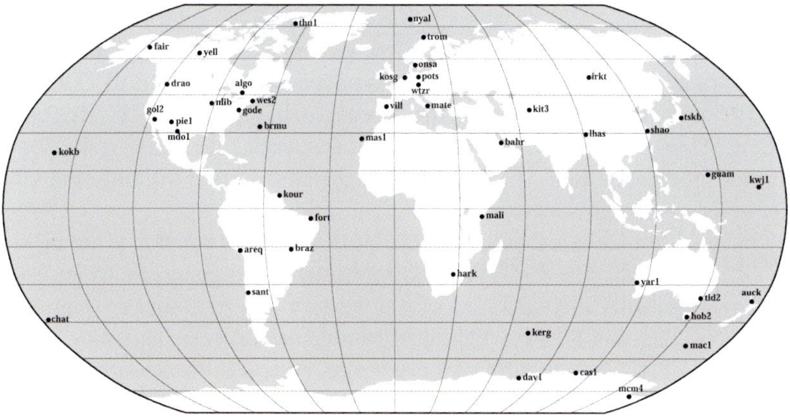

FIGURE 1 The IGS stations used in the CTRF establishment.

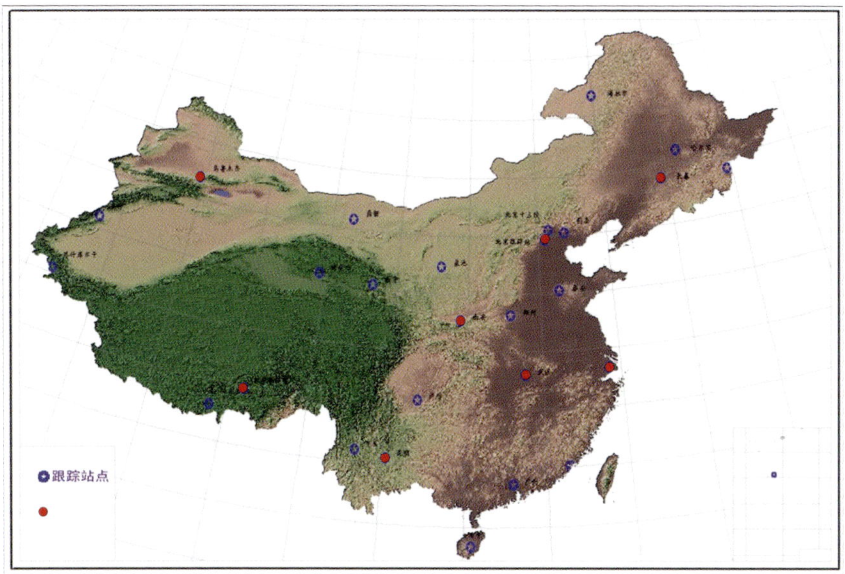

FIGURE 2 IGS stations in China.

Problems of CTRF 2000 and BTRF

Yet, there is quite some room for improvement for CTRF 2000 and BDC. The accuracy of CTRF 2000 is not so homogeneous compared to ITRF. Only 28 CORS and about 1,000 monitoring stations with high accuracy were used in the CGCS2000. The geometry of CTRF is quite weak. The positional velocities were not provided. BRTF is not integrated in the CTRF 2000; only the GNSS observation campaign was carried out.

One more problem is that the crustal deformation information is not included in CTRF or BTRF. The CRTF station movements are generally in an eastern direction. Especially the velocities of the western part of China are significant. Therefore, CTRF and BTRF, which were established 10 years ago, may be problematic in accuracy and reliability for the users with highly accurate positioning requirements.

Future of CTRF 2000 and BDC

CORS stations will be increased from 28 to 260 and will consist of 3 co-located stations with VLBI, six co-located stations with fixed SLR, and the establishment of 232 new stations. Regional stations will be increased from 1,000 to 2,000. Many more national CORS stations will be established and integrated.

BTRF will be connected to CTRF and ITRF by a new measurement project. For interoperability, setting up new overseas BTRF stations is in consideration. Multi GNSS (including BeiDou system) with their multiple frequencies should be applied in BTRF maintenance to realize interchangeability.

The new integrated adjustment of BTRF stations and updated CTRF stations will be carried out. The adjustment of the measurements of BTRF stations with the prior information of ITRF will be performed.

Unified orbit determination strategies using the same terrestrial tracking stations with the same measurements for all GNSS constellations are also beneficial for interchangeability.

BEIDOU TIME SYSTEM (BDT)

Definition of BDT

BDT is an internal, continuous navigation time scale, without leap second. The basic unit is the SI second. The largest unit used to state BDT is one week, which is defined as 604,800 seconds. BDT is counted by the week number (WN) and the second of week (SoW). The zero point is at January 1, 2006 (Sunday) UTC 00h00m00s.

Maintenance of BDT

BDT is realized by composite clocks. It is maintained by a time and frequency system (TFS) located at the master control station (MCS). TFS is mainly composed of five elements: (1) clock set (CS), (2) inter-measurement element (IME), (3) outer-comparison element (OCE), (4) data processing element (DPE), (5) and signal generation element (SGE). IME measures the original time and frequency signals from CS and gives out the clock differences both in time and frequency in a circular pattern. The offsets of BDT with respect to other time scales, especially to UTC, are obtained by OCE. DPE completes the calculation in a given algorithm to give a relative uniform time scale as the time reference for the whole navigation system, which is called BDT. In line with BDT, SGE exerts a frequency adjustment to the frequency signal from MC, and generates all the real physical time and frequency signals requested in MCS. In order to be as consistent as possible with UTC, BDT may steer to an interposed frequency adjustment after a period of time (more than 30 days) according to the situation, but the quantity of the adjustment is not allowed to be more than 5 × 10E-15.

Performance Analysis of BDT

The performance of BDT includes the evaluations of the time frequency accuracy, stability, and offset with UTC. The evaluation methods are usually divided into GPS satellite common view links (Liu et al., 2011), GEO satellite common view links, and two-way time frequency comparison links (Liu et al., 2009).

The test shows that the results obtained from the three kinds of links are consistent. Only the results obtained by GPS satellite common view link are given. The performance test was carried out from June 12 to October 8, 2010.

The bias of a hydrogen clock referred to BDT is shown in Figure 3. The bias deducted with frequency offset is shown in Figure 4. The frequency stability of the clock is shown in Figure 5. The time offset between BDT and NUTC is shown in Figure 6. The time offsets between BDT to UTC and BDT to GPST are also calculated through National UTC from NTSC (National Timing Service Center) and are shown in Figures 7 and Figure 8, respectively.

The performance of TFS is listed as follows: time accuracy is better than 2 × 10E-14; long stability is better than 1 × 10E-14/1 day, 6 × 10E-15/5 days, 5 × 10E-15/10 days and 6 × 10E-15/30 days. The time bias |BDT-UTC| is smaller than 100 ns (modulo one second).

The performance of the satellite clock is evaluated. The observed deviations of the satellite clock with respect to BDT and the bias of satellite clock taken off the mean clock rate are shown in Figures 9 and 10, respectively.

The observed Allen variance and Hadamard variance of the satellite clock are calculated and shown in Figures 11 and 12, respectively.

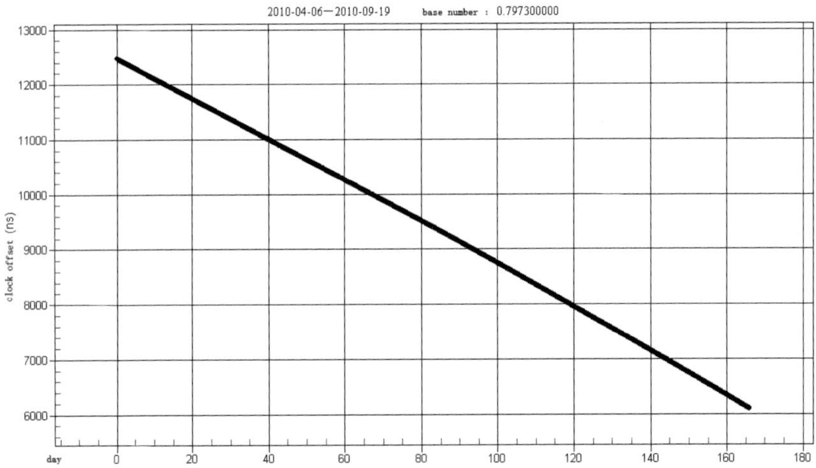

FIGURE 3 Bias of a hydrogen clock referred to BDT.

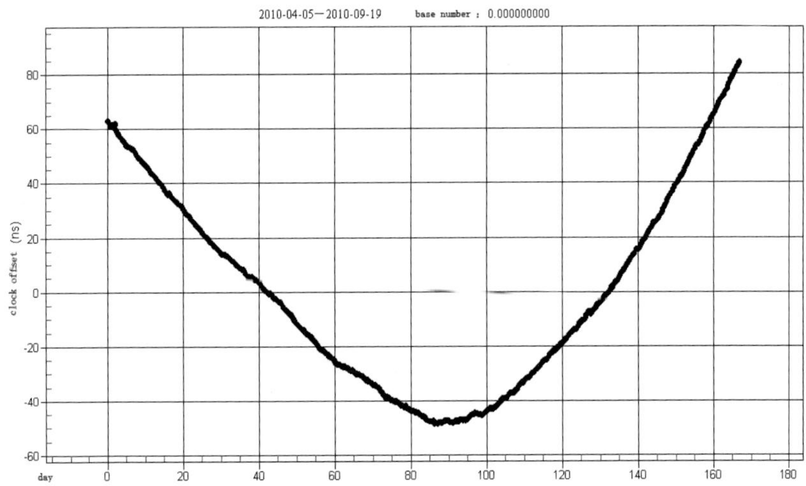

FIGURE 4 Bias deducted with frequency offset.

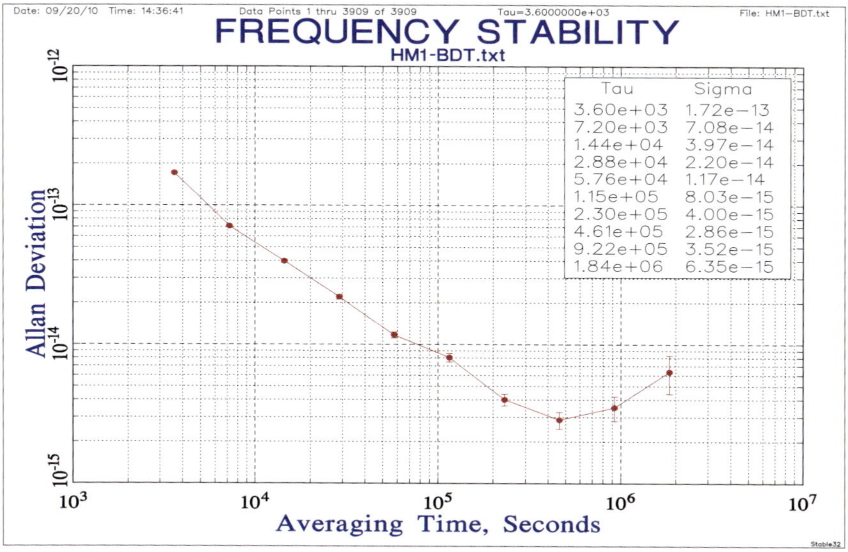

FIGURE 5 Frequency stability of the clock.

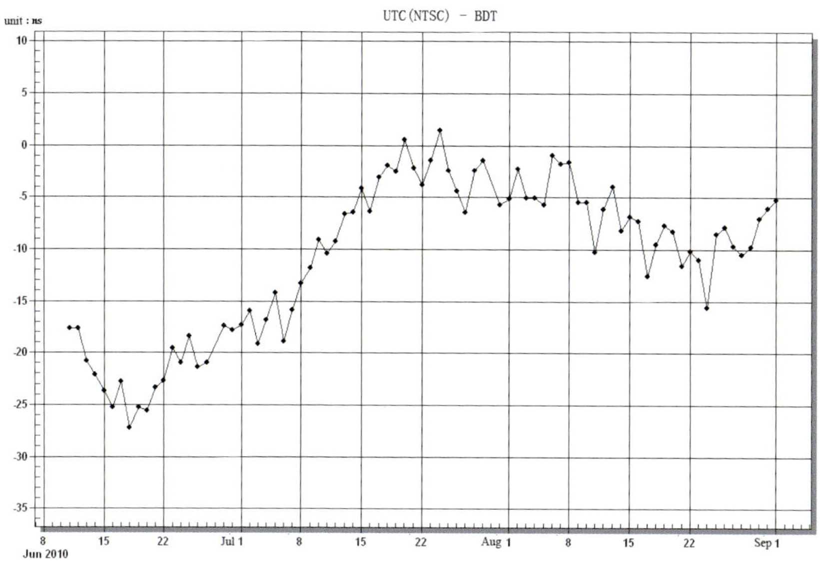

FIGURE 6 Time offset between BDT and NUTC.

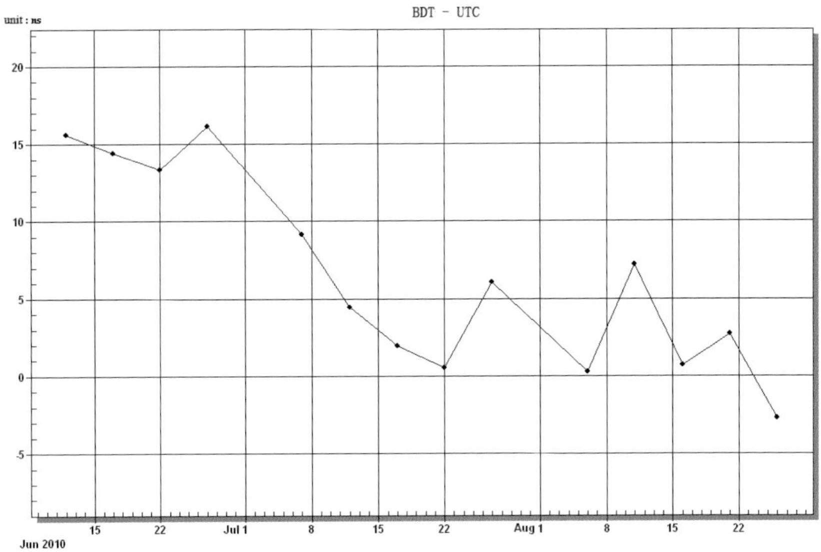

FIGURE 7 Time offset of BDT and UTC.

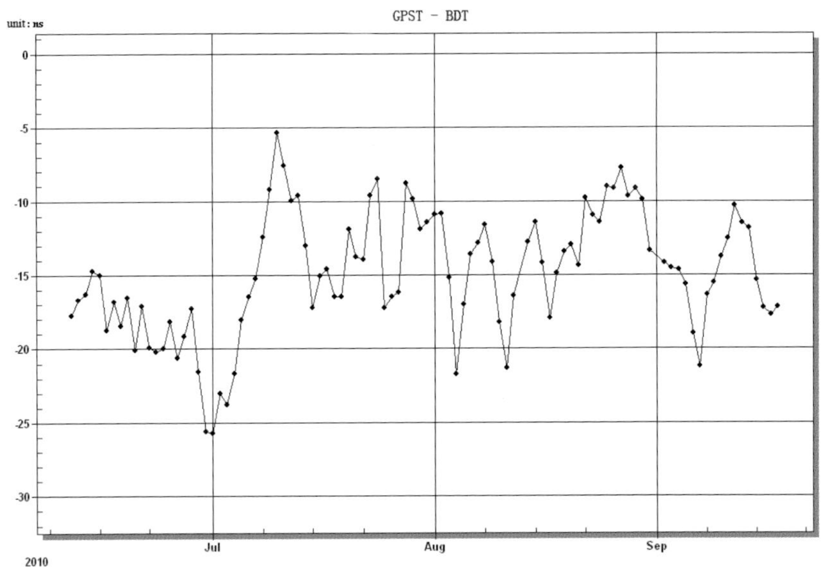

FIGURE 8 Time offset between BDT and GPST.

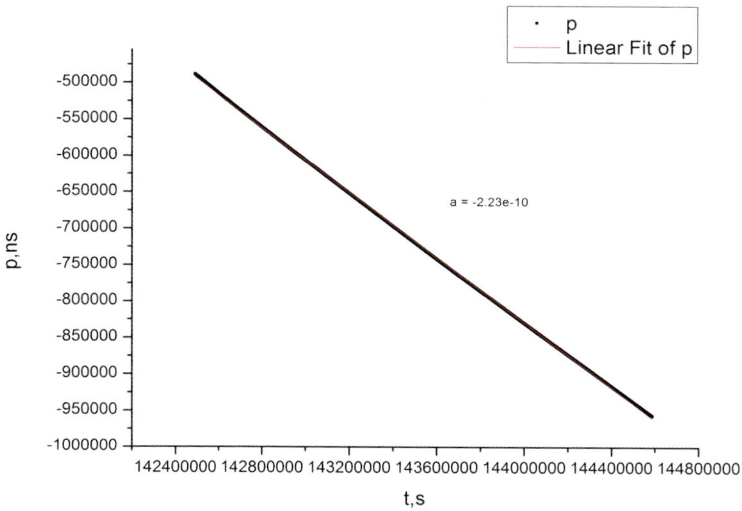

FIGURE 9 Bias of satellite clock with respect to BDT.

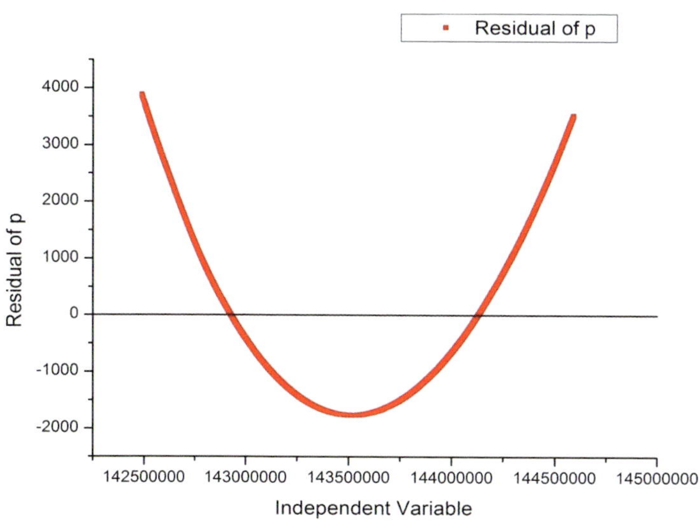

FIGURE 10 Bias taken off mean clock rate.

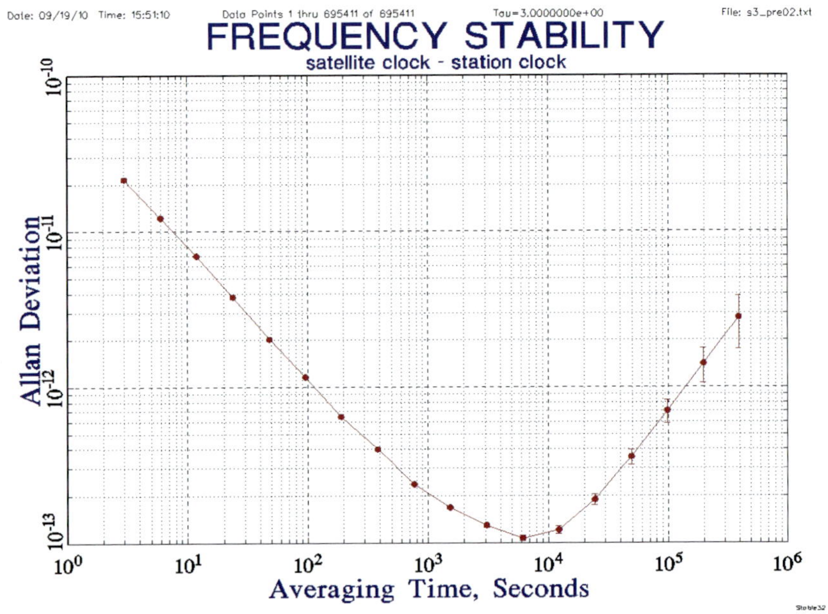

FIGURE 11 Observed Allan variance of satellite clock.

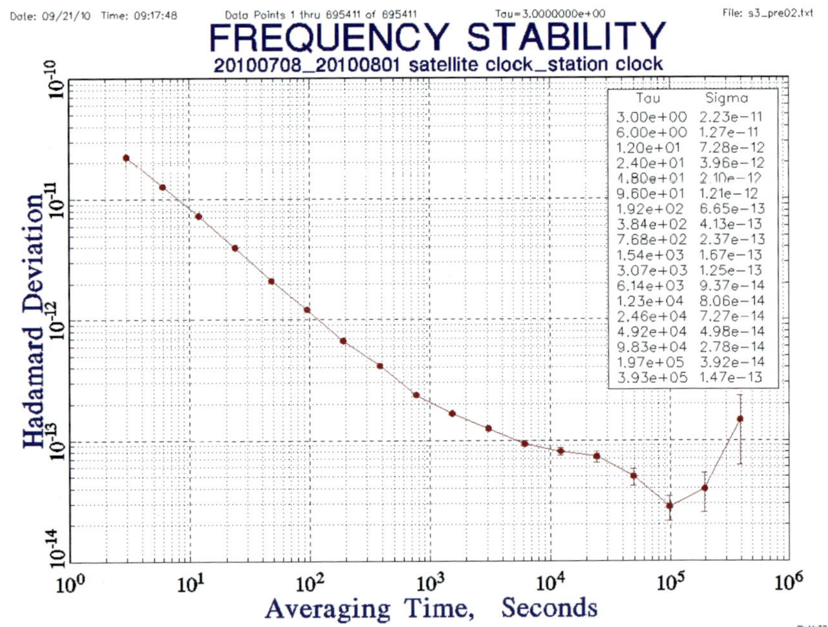

FIGURE 12 Hadamard variance of satellite clock.

The services of BDT are divided into one-way RDSS services, two-way RDSS services, and one way RNSS services. The uncertainty of RDSS one-way service is from 100 ns to 50 ns; the uncertainty of two-way RDSS service is from 20 ns to 10 ns; and the uncertainty of RNSS one-way service is about 50 ns.

Future Developments of BDT

First of all, the establishment of accurate relation/time offset between BDT to UTC needs to be strengthened. Two-way satellite time and frequency transfer is still working. There will build fiber chains of time and frequency transfer between MCS and the NTSC of the Chinese Academy of Science, National Institute of Metrology, etc. The time offset of BDT with respect to UTC can be obtained indirectly.

In order to improve the long stability of BDT and also improve the interoperability of BDT with other satellite navigation time systems, the number of clocks needs to be increased. Measurement hardware and software should be modified. The GNSS time monitor system is to be set up for observing the time differences and calculating the time offset. The offset or interoperability parameters of the time system will be broadcasted in BD NAV data. Thus, the time system of BeiDou navigation system will be well aligned to UTC for interoperability. For further interchangeability, new clocks with high accuracy and reliability are necessary.

CONCLUSIONS

The BDC system is aligned to ITRS and other GNSS coordinate systems. CTRF will be updated by new observation projects. The consistency of the BDC system and ITRS will be strengthened. A GNSS observation campaign will be regularly carried out between BTRF and CTRF. BeiDou time is accurate, stable. and reliable. BDC and BDT systems meet the requirements of compatibility and interoperability. The ultimate goal in improving BDT and BTC is to realize the interchangeability, which is a very long way to go.

REFERENCES

Liu, L., L. Zhu, and C. Han. 2009. The model and experimental analysis of two-way satellite and ground time transfer. Astronomy Paper 50(2): 189–196.

Liu, L., et al. 2011. The Method and Experiment Analysis for Two-Way Satellite Common-View Time Transfer. China Satellite Navigation Conference 2011 (CSNC 2011), Shanghai, China (in Chinese).

Wei. Z. 2003. National geodetic coordinate system: to next generation. Geomatics and Information Science of Wuhan University 28(2): 138–143 (in Chinese).

Yang, Y. 2009. Chinese geodetic coordinate system 2000. Chinese Science Bulletin 54: 2714–2721.

Yang, Y., M. Zha, L. Song, Z. Wei, Z. Wang, G. Ouyang, B. Xu, X. Wu, and J. Wang. 2005. Combined adjustment project of national astronomical geodetic networks and 2000' GPS control network. Progress in Natural Science 15(5): 435–441.

Yang. Y., Y. Tang, C. Chen, M. Wang, P. Zhang, X. Wang, L. Song, and Z. Zhang. 2007. National 2000' GPS control network of China. Progress in Natural Science 17(8): 983–987.

Yang. Y., Y. Tang, C. Chen, M. Wang, P. Zhang, X. Wang, L. Song, and Z. Zhang. 2009. Integrated adjustment of Chinese 2000' GPS control network. Survey Review 41(313): 226–237.

A Global Safety of Life Service from Multiple GNSS Constellations

PER ENGE
Stanford University

ABSTRACT

The Global Positioning System serves 1 billion civil users with applications including: navigation for cars, aircraft, ships, spacecraft, pedestrians, and emergency services; time transfer for telecommunications, finance, and power delivery; and a breadth of scientific uses. Cognizant of this utility, the Russians are rejuvenating their satellite navigation system, and new systems are being fielded by China, Europe, India, and Japan. Taken together, this Global Navigation Satellite System will provide geometric diversity with more than 100 navigation satellites, all with triple frequency diversity for civil use. This paper will briefly describe the technology with a focus on safety of life applications.

INTRODUCTION

Satellite navigation serves a wide breadth of applications based on the satellites depicted in Figure 1. As shown, the vast majority of these satellites are in medium Earth orbit (MEO), with some assistance from satellites in geostationary orbits (GEO). In 2011, most of these satellites belong to the Global Positioning System (GPS) originally developed by the U.S. Department of Defense. This development began in the 1970s when the planners predicted that GPS would serve a total of 40,000 military users with some ancillary civil use. Today, the civil community ships more than 10 million GPS receivers per month. The GPS tail now wags the dog, and the civil community has generated an unexpected breadth of applications.

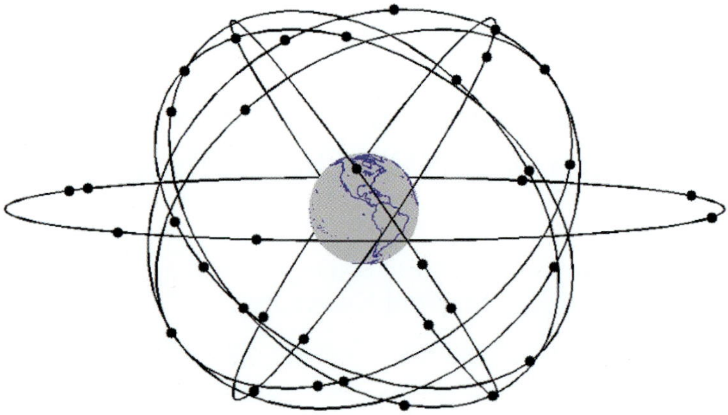

FIGURE 1 Today, GPS has approximately 31 satellites in medium Earth orbit. GPS is augmented by seven satellites in geostationary orbit. Source: Adapted from van Diggelen, 2009. Reprinted courtesy Frank van Diggelen.

For example, every new Boeing or Airbus aircraft carries a GPS receiver for navigation in the enroute and terminal area airspace. GPS is also used to guide aircraft while approaching airports. In some cases, it provides the critical vertical dimension of location down to altitudes of 200 feet.

In addition, most new smart phones or feature phones carry GPS receivers that have a bill of materials around $1. These receivers are used to guide our walking and driving lives. They also provide our location automatically to emergency services when we make such a call. In time, they will provide our location to good Samaritans standing next to automatic electronic defibrillators when someone suffers a heart attack. GPS receivers in cell phones will also serve significantly less uplifting applications that push advertising to our phones based on our location.

Other applications include ship navigation, pointing information for spacecraft, land survey, energy exploration; and time transfer for telecommunications, power delivery, and financial transactions.

Attracted by this utility, Russia is rejuvenating its satellite navigation system, called GLONASS. China is expanding their regional system, BeiDou (also known as Compass), to include global coverage. Europe has launched its first prototype satellites for the Galileo system. Japan and India have launched satellites for the regional systems. Figure 2 depicts a satellite navigation future based on more than 100 satellites from this family of systems. Taken together, we will have a Global Navigation Satellite System (GNSS).

The next section describes the new signals to be delivered by this multiplicity of constellations. It will be followed by a focus on my area of greatest interest:

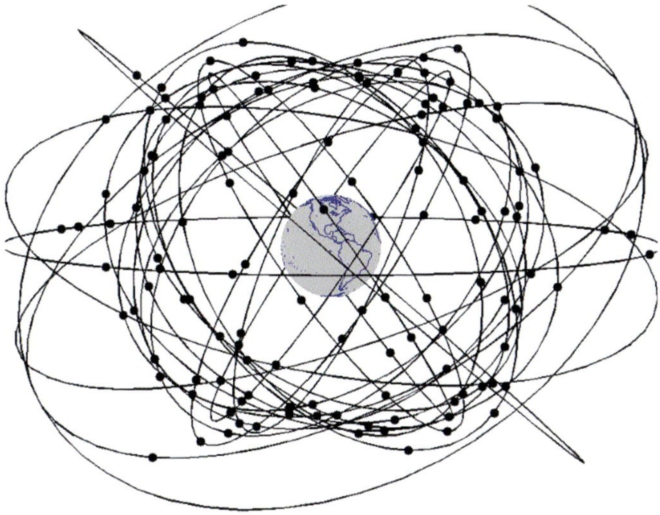

FIGURE 2 In the future, GNSS may grow to include more than 100 satellites, mostly in medium Earth orbit, with some in geostationary and inclined elliptical orbits. Source: Adapted from van Diggelen, 2009. Reprinted courtesy Frank van Diggelen.

safety of life applications and civil aviation. The following section discusses our continuing concern with radio frequency interference (RFI), and the final section provides a brief summary of the paper.

FREQUENCY DIVERSITY FOR SATELLITE NAVIGATION

The multiplicity of satellites described above will provide geometric diversity. Happily, the new satellites will also provide frequency diversity for civil users. Each new satellite will radiate civil signals at three frequencies rather than the single civil frequency offered today.

The top trace in Figure 3 shows the spectrum for the GPS satellites that are currently being launched. These new satellites broadcast at three civil frequencies: L1, L2, and L5. L1 is 1575.42 MHz and home for the so-called clear access (C/A) signal. This signal is a spread spectrum with a modest chipping rate of 1 Mcps. Even so, it is the basis for all of the civil applications described above. L1 C/A overlays military signals in this same band. L2 is 1227.60 MHz and also carries a civil signal with a 1 Mcps chipping rate on the seven most recent GPS satellites. Taken together, L1 and L2 provide redundancy to combat accidental RFI and a means to remove the dispersive delay due to the ionosphere. Both features are important. RFI is becoming more prevalent in the GPS bands, and the ionosphere is the largest natural error source.

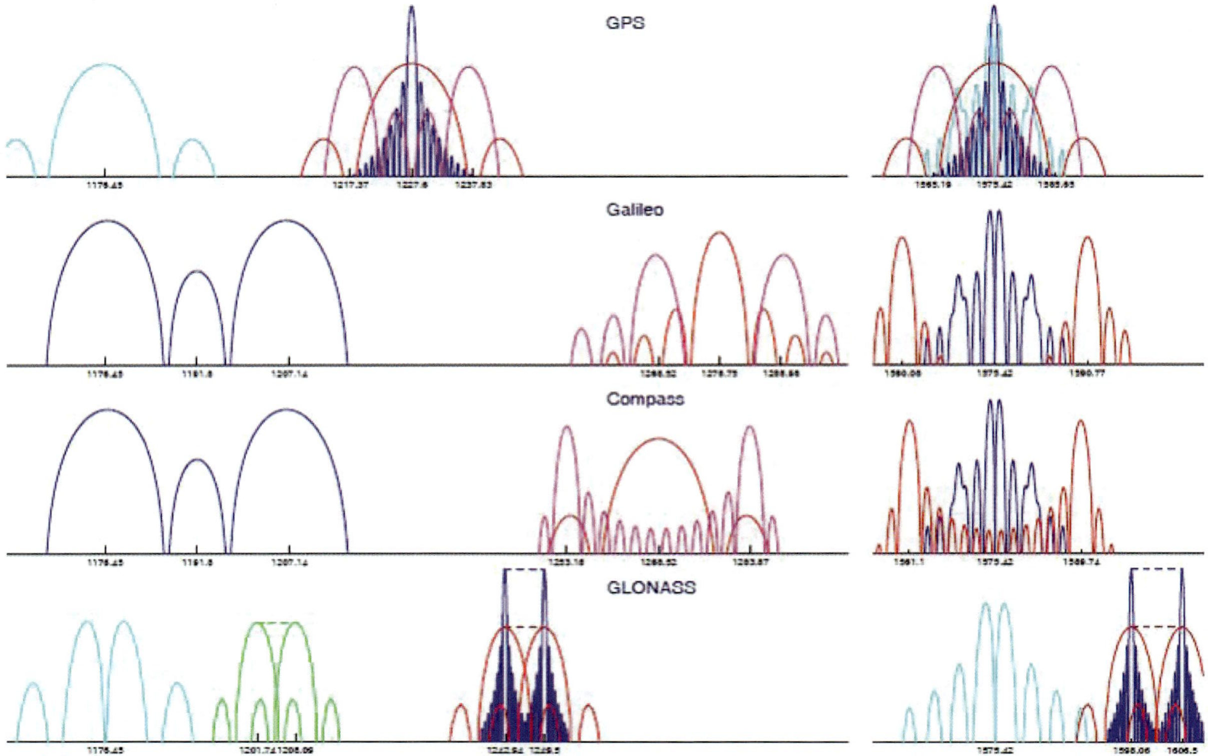

FIGURE 3 Signal spectra for GPS, Galileo, Compass, and GLONASS. From the left, new GPS satellites radiate at L5 (1176.45 MHz), L2 (1227.60 MHz), and L1 (1575.42 MHz). Source: Liang Heng, Stanford University. Reprinted with permission.

At 1176.45 MHz, L5 is the home for the first civil signal with a 10 Mcps chipping rate. It provides 10 dB more processing gain against RFI and can be used with L1 to remove the ionospheric delay. L1 and L5 are particularly important to aviation and other safety of life applications. They both fall in Aeronautical Radio Navigation System (ARNS) portions of the radio spectrum. Unlike L2, these ARNS signals may be used by civil aircraft to provide navigation in poor visibility as part of instrument flight rules (IFRs). Taken as a triplet, L1/L2/L5 allow receivers to form beat frequency signals that can be used to resolve the carrier phase ambiguity and enable phase interferometric measurements for very high precision.

The signals for GLONASS, Galileo, and Compass are also shown in Figure 3. As shown, they are not identical to the GPS signal shown in the top trace. However, they share the main features: triple frequency diversity with at least two signals in ARNS bands surrounding L1 and L5.

IMPACT ON CIVIL AVIATION AND SAFETY OF LIFE APPLICATIONS

Civil aviation has augmented GPS to detect faults or rare normal conditions that threaten flight safety. Indeed, most safety-of-life applications require mechanisms to mitigate the feared events that may present hazardously misleading information to the navigator. Faults are failures in the man-made portions of our navigation systems. The probability of a fault in the GPS system is approximately 10^{-5} per hour per satellite. The target level of safety for an aircraft navigation system is approximately 10^{-7} per hour or 100 times smaller than the observed failure rate. Similarly, rare normal conditions, notably the ionosphere, can introduce location errors that may be potentially hazardous to aircraft safety. These can occur several times per year during the peak of the solar cycle. In these peak years, the rate of these events is more than 100 times greater than the target level of safety for aviation. Hence, the civil aviation community has augmented GPS with systems that detect and remove errors due to these faults and rare normal conditions. Three such augmentation strategies exist.

(1) Ground-based augmentation systems (GBAS) are located at the airport to be served. Reference receivers monitor the GPS (or GNSS) signals. Because the reference receivers are at known locations, they can generate corrections to remove the nominal GPS errors, and they can generate alarms to flag satellites that cannot be reasonably corrected and characterized. All the reference receivers are on the airport property. Thus, the corrections and alarms are only valid within 100 kilometers or so around the airport. We call this the terminal airspace. Given this limitation on range of applicability, GBAS uses a VHF data broadcast to communicate with nearby aircraft.

(2) Space-based augmentation systems (SBAS) spread their reference receivers across continental areas. For example, some 38 stations are used to cover North America. These receivers send their GPS measurement data to master stations that generate corrections and error bounding data that is valid over the area spanned by the reference network. Because the data is valid over continental areas, SBAS uses a geostationary satellite to broadcast this navigation safety data to the airborne fleet.

(3) In contrast to SBAS and GBAS, aircraft-based augmentation systems (ABAS) are self contained. In fact, ABAS include a family of fault detection techniques known as receiver autonomous integrity monitoring (RAIM). SBAS and GBAS detect faults by comparing the GPS measurements to ground truth. RAIM compares the GPS measurement from one satellite to the consensus of the other satellites in view. Mathematically, RAIM is based on the residuals of the individual GPS measurements relative to the least squares navigation solution based on all satellites in view. RAIM is attractive because it does not need a ground reference network or a real-time broadcast from the ground network to the aircraft. However, the RAIM fault detection capability is weak compared to SBAS or GBAS because it does not have access to ground truth. The navigation solution must be over-specified and the geometries of the underlying subset solutions must be strong. Thus, RAIM has not yet been used for vertical guidance. It has only been approved for lateral guidance.

With the advent of the new constellations described above, RAIM may be able to support vertical navigation. After all, geometric diversity means that the navigation solutions will be over-specified and that the subset geometries will be stronger. The air navigation community is researching this possibility and has developed a concept imaginatively known as advanced RAIM or ARAIM.

Like its predecessor, ARAIM is based on the high availability of over-specified navigation solutions. However, it is subject to a more stringent safety proof than RAIM, because ARAIM targets vertical guidance and must overbound the vertical errors at the 10 m level. In contrast, RAIM for lateral guidance only needed to protect against lateral errors of 200 m or more. For these reasons, ARAIM has been subject to greater scrutiny than RAIM, and this examination has found an issue of concern: the safety community is concerned that ARAIM would not detect any faults that caused the entire GNSS constellation to rotate relative to the Earth. These would not be detected by a residuals test but could be hazardous. After all, the aircraft would have an error relative to the Earth's surface. Hence, current research focuses on architectures that would combine ARAIM and SBAS or GBAS.

If a combination of ARAIM and SBAS can be proven to be safe, then it may be able to support navigation down to altitudes of only 100 feet over the airport surface. Because ARAIM + SBAS would be a multi-constellation capability, it would be independent of the health of any one of the core GNSS constellations. Aviation and safety-of-life applications would welcome this robustness.

RADIO FREQUENCY INTERFERENCE

As mentioned earlier, GPS satellites broadcast from MEO, and so the satellites are approximately 12,000 miles above the receivers. These satellites are placed at this altitude so that an individual satellite covers one-third of Earth's surface. With 30 satellites carefully arranged in MEO, all earthbound users of GPS (with a clear view of the sky) can see at least the prerequisite four satellites to instantaneously determine three dimensions of location plus time. MEO is used so that a reasonably sized constellation can enable navigation worldwide.

However, MEO signals are weak when they reach Earth, and GNSS signals have a received power of only 10^{-16} Watts. They can be easily overwhelmed by Earth-sourced interfering transmissions at the GPS frequency. This RFI can be *scheduled*, *accidental*, or *malevolent*. Amongst these, deliberate interference, called jamming, is the looming threat to safety-of-life applications.

In the past year, so-called personal privacy devices (PPDs) have become widely available on the Internet, and such a jammer is shown in Figure 4. The

FIGURE 4 Example of a personal privacy device (PPD).

most inexpensive PPDs are single antenna devices that jam the one GPS signal frequency (L1) that is used by most users. More expensive units have multiple antennas and attack all three GPS signal frequencies (L1, L2, and L5). As such, these attackers anticipate the next generation of GPS user equipment that would continue to function if only one or two of the three frequencies were jammed. Other PPDs jam GPS and the cell phone frequencies at the same time, shutting down all calls. They are preferred by car thieves that wish to prevent on-car warning systems from reporting the GPS location of a stolen car to the authorities.

PPDs range in price from $30 to more than $300 based on the number of frequencies under attack and the transmitted power. They radiate powers from a few milli-watts to hundreds of milli-watts. The former knock out GPS receivers for hundreds of yards, and the latter can have dangerous effects for many miles.

As their name suggests, PPDs are marketed to individuals that fear for their privacy. This sales strategy seems to be effective. An investigation recently initiated by the Federal Aviation Administration (FAA) revealed that trucks traveling on the New Jersey Turnpike carry these devices. Perhaps, the drivers worry that the company dispatcher is monitoring their locations. Ironically, the attention of the dispatcher must be drawn to the truck that never provides location reports.

In any event, a PPD can cause collateral damage much greater than any privacy protection the user may possible enjoy. The above-mentioned FAA investigation was sparked while a GBAS at Newark International Airport was being installed. As described above, GBAS uses GPS receivers at the airport to aid GPS receivers in the approaching aircraft. The GBAS antennas were placed next to the New Jersey Turnpike, and the ground receivers suffered frequent continuity breaks during the test period. These breaks were due to PPDs carried by vehicles on the New Jersey Turnpike. The United States is currently investigating legal and technical remedies for PPDs. At Newark, the antennas will be relocated, and the system will be reconfigured to better tolerate short outages on a subset of the antennas.

For the longer term, the FAA focuses on alternate position navigation and time (APNT) based on terrestrial radio navigation and inertial navigation. Today, the aviation community continues to operate the full set of terrestrial navigation aids that predated GPS. These include VHF omni-directional range (VOR), distance measuring equipment (DME), and instrument landing systems (ILS). If GPS is jammed or suffers from some other systemic failure, pilots can revert to these systems. However, this terrestrial network is expensive and does not support the navigation capabilities needed for the Next Generation Air Traffic Management System (NextGen). It needs to be reconfigured to reduce cost and provide area navigation rather than point-to-point navigation. NextGen is predicated on area navigation, which is not fully backed up by today's ground-based network.

To this end, the APNT effort focuses on the provision of area navigation for all enroute and terminal airspace over CONUS and approach navigation for all airports required for safety and commerce. APNT should be independent of GNSS. Rather it should be based on navigation aids that can be cost effectively

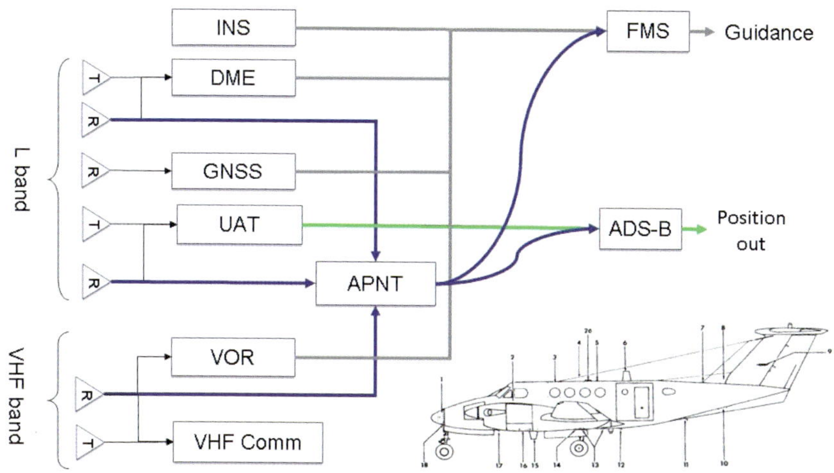

FIGURE 5 Potential sources of alternate position, navigation and time, including inertial navigation systems, distance measuring equipment, universal access transponder, VHF omni-directional radio, and VHF communications.

retained or are planned for installation in the next decade. This investigation has included all of the navigation aids shown in Figure 5 and has focused on DME and ground-based transmitters planned for the next generation of surveillance.

SUMMARY

GPS serves well with approximately 1 billion users utilizing a surprising breadth of applications. In the future, the capability of satellite navigation will multiply as GPS is joined by a rejuvenated system from Russia and new systems from China, Europe, India, and Japan. This multiplicity of ranging sources will provide geometric diversity. Moreover, each of these satellites will radiate ranging signals for civilians at three or more frequencies, and some of these new signals will have 10 times the bandwidth of the GPS C/A code signal that enables the vast majority of present applications. Thus, we can expect a new wave of applications that leverage these new technical capabilities.

However, radio frequency interference will be a continuing challenge to those applications that involve safety of life. Concern for privacy is great, so jammers designed to protect privacy will wreak collateral havoc. Moreover, the pressure to find new radio spectrum for worldwide Internet access will endanger the weak GNSS signals coming from medium Earth orbit.

REFERENCE

van Diggelen, F. 2009. A-GPS: Assisted GPS, GNSS, and SBAS. Norwood, Mass.: Artech House.

Monitoring and Assessment of GNSS Open Services

JIAO WENHAI
Beijing Institute of Tracking and Telecommunication Technology

DING QUN
Xi'an Research Institute of Navigation

LI JIAN-WEN
Zhengzhou Institute of Surveying and Mapping

LU XIAOCHUN
National Time Service Center

FENG LAIPING
Xi'an Research Institute of Surveying and Mapping

Interoperability of GNSS Open Services has already been a significant tendency in developing all satellite navigation systems, and its performance will directly affect the security and reliability of its usages. Therefore, monitoring and assessment of GNSS Open Services have become a focus of attention for all providers and users of GNSS. This paper begins to illuminate the elements and methods for monitoring and assessing GNSS Open Services. Then according to the requirements, the architecture of an international GNSS Monitoring and Assessment System (iGMAS) is designed to achieve the 4-overlap and 1-overlap coverage and sophisticated analysis, respectively. Here, this iGMAS is based on omnidirectional antennas, multi-beam antennas, and a high-gain paraboloid antenna. In the meantime, the configuration scheme of worldwide monitoring stations is provided. Finally, some related works that have been done to monitor and assess the BeiDou Open Service are introduced that can be used to verify the feasibility of this proposed system.

INTRODUCTION

Stepping into the 21st century, because of the successful launches of the BeiDou and the Galileo satellites, the GNSS is changing from the bipolar competition situation between GPS and GLONASS to a new situation of competition and cooperation among four systems. Obviously, the compatibility and interoperability of navigation signals have already been the main features during the period of GNSS development. Against this background, open signals, with the function of interoperability, are expected to bring GNSS services of higher quality and better performance to GNSS users, especially in such places as urban canyons and mountainous areas where visual airspace of satellite is limited. Open signals can bring a significant increase in the number of visible navigation satellites, improving the reliability and availability of navigation and positioning services (Yang, 2010). So far, the main existing and being designed signals of GNSS Open Services are as shown in Table 1.

At present, the four satellite navigation systems have issued or are planning to issue their own specifications on open service performance. However, because of the diversity in their respective conditions and knowledge, there may be large differences in those specifications, either in form or in performance. As a result, users will get confused when they are using and it will be inconvenient for them to use. In addition, there are no performance specifications on GNSS open service signals. To ensure the safety of the usage and achieve the ultimate goal of interoperability of GNSS Open Services signals, it is essential for us to research the monitoring and assessment of GNSS Open Services.

Monitoring and assessment of GNSS Open Services could provide third-party information on performance for a single system and reliable decision-support

TABLE 1 Signals of GNSS Open Services

Global Navigation Satellite Systems	Frequency	Center Frequency (MHz)	Modulation Mode	Interoperable or Not
GPS	L1 C/A	1575.42	BPSK(1)	
	L1C	1575.42	MBOC(6,1,1/11)	Yes
	L2C	1227.6	BPSK(1)	
	L5C	1176.45	QPSK	Yes
GLONASS	L1OF/L1OCM	1598.06~1604.40	BPSK	
	L2OF/L1OCM	1242.94~1248.63	BPSK	
	L3 OC	1202.025/1207.14	BPSK	
BeiDou	B1-C	1575.42	MBOC(6,1,1/11)	Yes
	B2a	1191.795	AltBOC(15,10)	Yes
	B2b			
GALILEO	E5a	1191.795	AltBOC(15,10)	Yes
	E5b			
	E1	1575.42	MBOC(6,1,1/11)	Yes

information for users when they are using its service, to minimize possible adverse effects and to improve quality of service. On the other hand, based on compatibility and interoperability, the GNSS system can bring not only a great leap in improved performance and quality of navigation service, but also large degradation due to signal interference problems between systems. Carrying on the monitoring and assessment of GNSS Open Services can help to deal with these problems. It can enhance the reliability of open service on one hand, and, on the other hand, from the standpoint of system monitoring and assessment on the comprehensive properties of multi-system open services, could improve compatibility and interoperability performance between each system, providing decision-support information for maintenance and management of the multi-system.

Since 1997, Stanford University has been under contract to support the U.S. GPS Joint Program Office by regularly monitoring the L-band transmissions of all newly launched and currently operational GPS satellites. The goal of this program is to verify that all deployed satellites are consistently and reliably performing according to the specifications. Stanford University has been measuring the signal power, code delay, frequency content, and bandwidth to verify the proper operation of satellite constellations. Currently, Stanford University is upgrading the RF and data-collection systems at a 47 m- (150 ft-) diameter steerable antenna ("The Dish") that has been used to collect these measurements. Meanwhile, it can also monitor and analyze other countries' satellite navigation systems.

In September 2005, the Institute of Communications and Navigation of the German Aerospace Center (DLR) established an independent monitoring station for analyzing GNSS signals. The core of this facility is a 30-m deep space antenna located at DLR ground station at Weilheim, Germany. The integrated measurement system fulfills the highest quality standards to obtain high-accuracy measurements from raw data of GNSS signals to perform precise analyses. After the commissioning phase of SVN49, relative data were collected in order to provide a basis for analyzing the signal anomaly. After the SVN49 signal anomaly was first noticed by this monitoring station, they started a detailed investigation of this issue using the high-gain antenna.

Stanford and the DLR have already formed the cooperation mechanism for joint monitoring and assessment of GPS and Galileo signals.

ELEMENTS AND METHODS FOR MONITORING AND ASSESSING GNSS OPEN SERVICES

The work of monitoring and assessing of GNSS Open Services can be divided into the following layers: constellation state layer, spatial signals layer, navigation information layer, and service performance layer. Before choosing and determining the elements and methods of assessment, we should, in the sight of different users' requirements, take into comprehensive consideration organic connections between different layers, as well as characteristics of independent elements. So the independence of open services for a single system cannot be neglected in front of multi-systems. Those finally determined elements and methods should be open, extensible, and compatible. So when a new system and its open signals appear, this monitoring system can analyze them as well. In addition, the function and role of third-party monitoring cannot be fulfilled unless real-time monitoring and real-time release are emphasized when monitoring and assessing GNSS Open Services.

Constellation State

Constellation state is one of the decisive factors for satellite navigation systems to realize their service performance. So real-time monitoring and assessment play a significant role for the users to use open services safely and reliably. Monitoring and assessment of the constellation state include the following content.

Constellation State

Constellation state refers to the working state of a single satellite or that of each satellite in GNSS constellation at a particular moment. Common working states can be divided into normal working state, testing and maintenance state, fault state, etc. The number of satellites in various states and their orbit distributions are used to describe those working states. In addition, such features as the launch date and service life of each satellite can also be used as a reference in assessing satellite state. Besides, information on constellation malfunction, maintenance, and management released alone by each satellite navigation system is also an important source of information for monitoring and assessing constellation conditions.

Constellation DOP

Constellation DOP indicates values of various DOPs of healthy satellites in their service range in a single or GNSS constellation, including GDOP, PDOP, HDOP, VDOP, TDOP, etc. As an influence factor of satellites geometry distribution on the error of navigation, positioning, and timing, DOP can be a comprehensive reflection of the geometry distribution and health condition of constellations. As for single-system constellations, instantaneous DOP is the most direct indicator for assessing performances. When it comes to multi-system constellations, instantaneous DOP can also be used as an important reference indicator for assessing the navigation and positioning performance of interoperable open signals.

Number of Visible Satellites

The number of visible satellites is still an important indicator for assessing the space-time condition of constellations. Its instantaneous value has a strong dependence on instantaneous DOP, whether it is a constellation of a single system or multi-systems. As the interoperability of GNSS Open Services comes true, the availability of more visible satellites will definitely reduce the sensitivity of a single satellite to service performance and improve observations redundancy at the same time. As a result, the precision, reliability, and availability of their services can be improved greatly.

Spatial Signals

To assess the quality of GNSS spatial signals, relative assessment experiments should be taken from time domain, frequency domain, modulation domain, and correlation domain.

Time Domain Characteristics

To obtain baseband signals, those digital intermediate frequency signals collected by a high-gain antenna should be conducted by quadrature carrier stripping and Doppler removing using a software receiver. Then we can draw time domain waveforms. And based on these, some characteristics such as the edge shape of one chip, code sequence, code rate, code shape, degree of digital distortion and analog distortion, etc. could be analyzed in detail (Hegarty and Ross, 2010). Besides, eye diagrams of actual signals and ideal signals could be displayed on the same graph to observe their similarity. In the meantime, related parameters of eye diagrams, for example, open degree and noise tolerance, etc., can be calculated. Thus we can assess the quality of received signals from time domain properties.

Frequency Domain Characteristics

Frequency domain characteristics are analyzed mainly with offline analysis software, assisted by some standard measuring instruments such as a real-time spectrum analyzer and so on. Through the test on signals from such aspects as carrier frequency, power spectrum and its envelope, bandwidth and center frequency, beamwidth between two zero point on main lobe, etc., we can see the difference of power spectrums between actual signals and ideal signals. In addition, some other indicators such as spectrum asymmetry or distortion, signal stray, and carrier leakage, etc., can be assessed comprehensively. Thus we can assess the quality of received signals from frequency domain properties.

Modulation Domain Characteristics

Both signal vector diagram and histogram are analyzed in detail mainly with offline analysis software, assisted by some standard measuring instruments such as a vector signal analyzer and so on. Using offline analysis software, vector diagrams of received signals can be drawn, with that of ideal signals in the same diagram. Then we can easily compare these two vector diagrams. Besides, error vector diagram can be constructed to extract indicators such as the phase orthogonality, carrier orthogonality, amplitude imbalance of I/Q channels, etc. In this way, phase error and amplitude error caused by channel distortions and noise interference and so on could be analyzed. In addition, histograms drawn by offline analysis software can be used to assess noise level. Using a vector signal analyzer, a signal vector diagram can be drawn, with its center frequency offset, SNR, EVM value, etc. shown on the screen. We can also see its stability by root-mean-square deviation calculated by this analyzer.

Correlation Domain Characteristics

Correlation domain characteristics of received signals are analyzed mainly from aspects such as the correlation curve, correlation loss, curve symmetry, code delay, and S curve deviation of DLL discriminator, etc., by offline analysis software (Lu and Zhou, 2010). The correlation curve of received signals is compared with that of ideal duplicated code sequences to see its asymmetry and false lock probability. Thus we can assess the degree of correlation distortion caused by band limited filter, noise, multipath, cross-correlation, etc.

Navigation Information

Navigation information refers to those parameters such as satellite constellation, satellite clock correction, ionosphere delay, time system deviation, etc., that are provided by a satellite navigation system. It is the most important factor that affects system service performance and is one of the most important factors for monitoring and assessing GNSS Open Services signals.

Message Validity

The verification of navigation message validity includes the following two points—consistency verification and the rationality and validity verification. As for consistency verification, various parameters such as time information, constellation, satellite clock, ephemeris, correction information, and so on are verified to see if they are set and updated correctly according to the ICD file. In the period of rationality and validity verification, it is necessary to determine according to the specific attribute and generation method of each parameter.

Accuracy of Ionosphere Delay Correction

Model parameters of ionosphere delay correction are very important to single-frequency users. Those ionosphere models and parameters in each satellite navigation system are far from different. Accuracy of ionosphere delay correction has a close relation not only with model types but also with accuracy and update frequency of parameters. It can be assessed by obtained values of ionosphere delay using double-frequency or multi-frequency pseudorange observations.

Accuracy of GNSS Time Deviations

GNSS time deviation reflects the deviations among timing biases of each satellite navigation system. It is important basic data to realize GNSS compatibility and interoperability. And its accuracy will directly affect the PVT calculating precision for various GNSS user equipments, especially for timing equipments. So monitoring GNSS time deviations has more requirements for equipment to get more reliable external accuracy.

Validity of Deviation Parameters Between Different Frequencies

Monitoring of deviation parameters between different frequencies contains the following two parts. One is the time deviations of different frequency signals when passing through satellite-borne equipments; these deviations have something to do with the calibrated values and signal frequency. The second is the time deviations among different modulated codes for the same carrier, known as the phase consistency between carrier and pseudo-code.

Service Performance

Satellite navigation systems provide three kinds of basic services—positioning, velocity measuring, and timing (PVT)—for all types of users. Its service performance is commonly described by such indicators as precision, availability, integrity. and continuity (DOD, 2008).

Precision

The precision of spatial signals includes URE, URRE, URAE, and UTE. It is an important indicator to assess the effect of satellite ephemeris and forecasted clock deviation on the error of positioning, velocity measuring, or timing. To assess the precision of signals, it is necessary to calculate the precise orbits and clock deviation afterwards.

Service precision is used to describe the deviation between measured value and the ideal value of positioning, velocity measuring, or timing for system users. It equals UERE multiplied by DOP, where UERE is determined by URE (User Ranging Error) and UEE (User Equipment Error). UERE can be calculated from the expression $UERE = \sqrt{(URE)^2 + (UEE)^2}$, and it is dependant on errors of satellite atomic clock, ephemeris, atmosphere model, multipath effect, receiver clock deviation, etc.

Availability

Service availability refers to the time percentage when the error of positioning, velocity measuring, or timing can meet the required threshold in a period of time (a return cycle) and in a specified service area. According to the expression that precision equals UERE multiplied by PDOP, service availability can be changed to DOP availability when setting UERE.

Integrity

Service integrity mainly refers to the probability for a navigation system to provide a timely alarm within the limited period when the error of its positioning, velocity measuring, or timing is larger than the threshold. It is generally expressed by alarm threshold, alarm time, and hazard misleading information (HMI), where alarm time refers to the maximum time delay allowed from the time a malfunction starts to the time an alarm generates. In addition, HMI probability refers to the dangerous probability when the current measured value falls in the range of alarm threshold.

Continuity

Service continuity refers to the time percentage when a satellite navigation system can continuously meet required service precision in a period of time and specified service area. The average continuity of service can be calculated according to MTBF (Mean Time between Failures) and MTTR (Mean Time to Repair) under the condition of the satellite's unplanned interrupt.

FRAMEWORK OF IGMAS

Architecture and Functions

To monitor and assess GNSS Open Services worldwide, it is very necessary to construct an iGMAS. The basic functions of iGMAS should include data monitoring and collecting, data transmitting, data storing, data analyzing, and information release. Figure 1 shows its architecture.

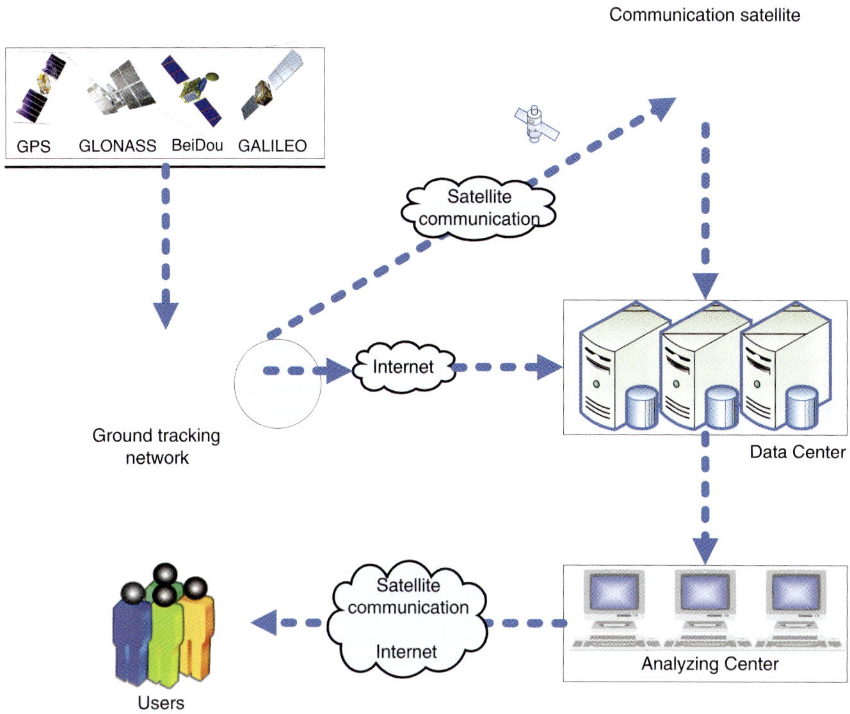

FIGURE 1 Framework of iGMAS.

Data Monitoring and Collecting

Data monitoring and collecting refers to the collecting and measuring of GNSS navigation signals, in addition to the collecting of environmental observation data (such as electromagnetic data, meteorological data). The main functions include:

1. Providing 4-overlap coverage observations for integrity monitoring and precise orbit determination.
2. Real-time monitoring on the quality of all GNSS satellite signals using digital multi-beam antennas and related equipment.
3. Realizing continuous and sophisticated observations on key satellites using a 30-m high-gain antenna and related equipment.

Equipment for monitoring and data collecting mainly includes anti-multipath omnidirectional antennas, multi-beam antennas that can receive all GNSS signals in the visible world (with equivalent diameter of 2.4 m and gain of 28 dB), a 30-m antenna, monitoring receiver, RF observations collecting equipment, baseband signal collecting equipment, delay calibrating equipment, meteorological sensing equipment, and electromagnetic environment analyzing equipment. Some other auxiliary equipment includes atomic clocks, data processor, network switches, etc.

Data Storing

Basic tasks of data storing include the classifying, storing, and management of monitored and collected data, and analyzing results, other data, or information, etc.

There are some data transmission links such as the Internet, satellite link (VSAT), and wireless mobile communication network, etc.

Data Analyzing and Information Release

Data analyzing and information release refers mainly to the classifying and processing of GNSS monitoring data. That means comprehensive analysis of GNSS signals and information to assess its service performance. For example, analyzing constellation characteristics such as the number of visible satellites, constellation state, and constellation DOP; analyzing navigation signal properties from time domain, frequency domain, modulation domain, and correlation domain; verifying the validity of navigation information, such as navigation message, accuracy of ionosphere delay correction, accuracy of GNSS time deviations, validity of time deviation parameters between different frequencies; and assessing the precision, availability, integrity, and continuity of GNSS Open Services. And, in the meantime, the release of statistical information on the working states of GNSS satellites to users based on those comprehensive analysis results. Here the equipment required includes real-time states displaying equipment, analyzing, and assessing software, servers, working stations, network exchanging equipment, mass storage, etc.

Data Transmission

Basic tasks of data transmission include data exchanging, control instruction transmission, and transmitting released information to users.

There are some data transmission links such as the Internet, satellite link (VSAT), and wireless mobile network, etc.

Tentative Plan of iGMAS Implementation

Preliminary Configuration Scheme

In iGMAS, those tracking stations scattered worldwide are used to implement data collecting. There are mainly two factors in considering their configuration:

1. To achieve signal-quality monitoring, it is necessary to use multi-beam antennas and a 30-m high-gain antenna to track satellites continuously. So at least those tracking stations should meet the requirement of 1-overlap coverage of satellites.
2. To achieve satellite-integrity monitoring, those tracking stations should meet the requirement of 4-overlap coverage of satellites.

Here the tracking stations are selected from existing IGS stations, and a grid-analysis method is used to optimize their layout (Stanton and Strother, 2007), with the analysis range between 65° N and 65° S and the height of 19,000 km. Those grids are divided into 1°*1° with the height angle's deadline of 15°.

To meet the two requirements described above, preliminary analysis results shows that 6 tracking stations and 24 stations are needed. In Table 2, we can see the locations of iGMAS tracking stations. Two kinds of optimal design methods for 1-overlap coverage and 4-overlap coverage are shown in Figures 2 and 3, respectively.

Operational Mode

Here the iGMAS can be constructed through international cooperation among providers and global civil users of the four global navigation systems.

The basic unit of iGMAS is called a node. Nodes can have five functions of data collecting and monitoring, data transmitting, data storing, data analyzing, and information release.

All nodes can be divided into three levels. Level one has all the five functions described above; level two has functions like data collecting, transmitting,

TABLE 2 Locations of iGMAS tracking stations

Name	Location	Name	Location
BEIJ*	China	MAS1	Spain
PERT*	Australia	BAHR	Bahrain
RCMN*	Kenya	ONSA	Sweden
LPGS*	Argentina	SANY	China
CAGS*	Canada	IRKT	Russia
TAH1*	France	WLMQ	China
CHAT	New Zealand	NOUM	France
KOKB	America	KERG	France
COSO	America	HART	South Africa
FAIR	America	CAS1	Antarctic
BOGT	Colombia	OHI3	Antarctic
FORT	Brazil	KELY	Greenland

* Denotes those tracking stations belonging to both 4-overlap and 1-overlap coverage.

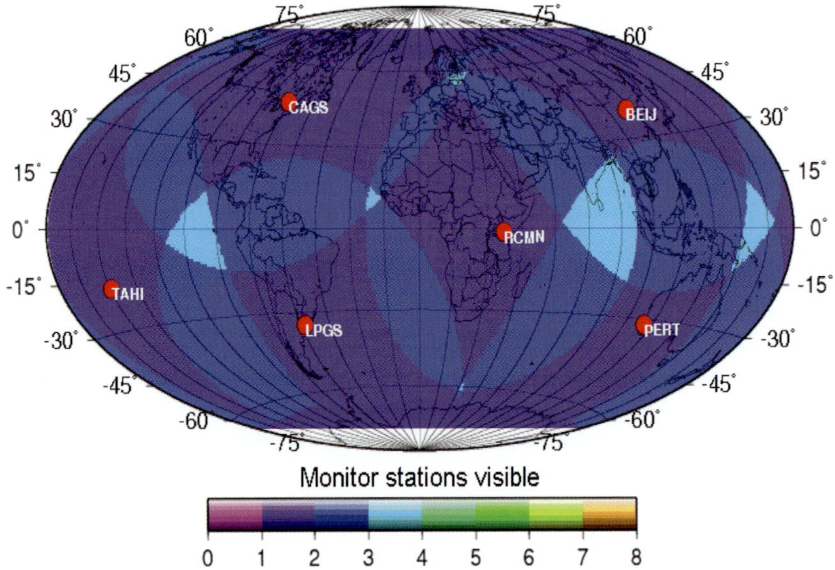

FIGURE 2 Optimal design of 1-overlap coverage stations.

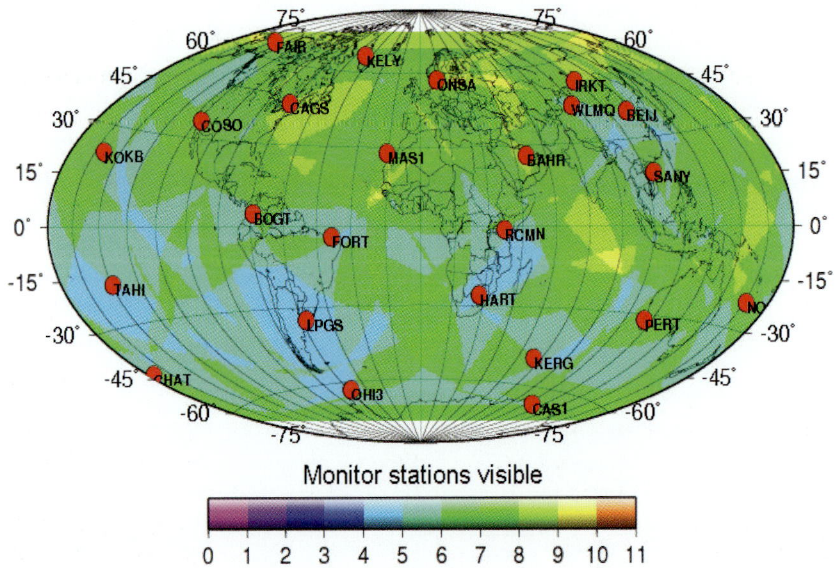

FIGURE 3 Optimal design of 4-overlap coverage stations.

storing, and analyzing; level three has only two functions of data collecting and transmitting. To realize the goal of data sharing, each node provides data to the net while automatically extracting data from it to perform signal assessing. All nodes in level one will join together to realize the monitoring and assessment of GNSS Open Services and information release. The net can work regularly through the collaborative work of its nodes, which can access or exit the net freely.

PERFORMANCE MONITORING AND ASSESSING FOR BEIDOU SYSTEM

Precision Monitoring and Assessing of Spatial Signals

To monitor and assess the performance of BeiDou open services from the information layer, a BeiDou monitoring and assessment system was built successfully in November 2007. It consisted of a monitoring and assessment center, one data analyzing center, and six tracking stations. The tracking stations were located in Xi'an, Shanghai, Changchun, Kunming, Urumqi, and the Antarctic, with each station equipped with BeiDou measuring receivers, high-precision atomic clocks, and computers, etc. The goal was to calculate the precise orbits and clock deviation of BeiDou navigation satellites, to assess the performance of satellite clocks and parameter accuracy of ionosphere models, and to verify the validity and rationality of navigation message, etc.

Quality Monitoring and Assessing of Spatial Signals

In February 2009, the first BeiDou signal-quality monitoring and assessment system was built successfully by NTSC, Chinese Academy of Sciences, in Lintong district of Xi'an, Shaanxi province. This system consisted of a 7.3-m antenna, an RF receiving subsystem, an RF observational data-collecting subsystem, a baseband signal-collecting subsystem, monitoring receivers, a calibrating subsystem, a data storing subsystem, and so on. Obtained results showed that this system could achieve some deeper tasks of GNSS signal-quality monitoring and assessment using many standard measuring instruments, monitoring receivers, high-speed data collecting equipments, and offline analysis software.

At present, two key technologies of GNSS signal-quality monitoring and assessment, that is, offline technology for analyzing signal performance based on correlation curve and channel calibrating technology, have strived to make important technological breakthroughs, while the anti-interference technology has also made much progress. Since its successful running in April 2009, this system has completed successfully signal-quality monitoring and assessment for BeiDou GEOs and IGSOs. In the meantime, it has also collected and analyzed GPS MEOs signals and Galileo Glove-B signals.

CONCLUSIONS

Monitoring and assessment of GNSS Open Services can bring much benefit to the healthy development of GNSS. It can not only extend the depth and breadth of GNSS applications, but also provide the foundation for all kinds of civil users to use system services safely and reliably. There is sufficient evidence to show that the iGMAS proposed in this paper is technically feasible. For example, the preliminary exploration of BeiDou in its open services monitoring and assessment, the long-term successful operation of IGS, and achievements in navigation signal monitoring and assessment made by Stanford University and DLR, etc. However, the construction of iGMAS is a global and long-term work, and it needs more organizations and countries to be involved. So it is very necessary to implement international cooperation.

REFERENCES

DOD (U.S. Department of Defense). 2008. Global Positioning System Standard Positioning Service Performance Standard. Available online at *http://pnt.gov/public/docs/2008/spsps2008.pdf*.

Hegarty, C.J., and J.T. Ross. 2010. Initial Results on Nominal GPS L5 Signal Quality. Pp. 935–942 in Proceedings of the 23rd International Technical Meeting of the Satellite Division of the Institute of Navigation (ION GNSS 2010), Portland, Oregon, September 2010.

Lu, X.C., and H.W. Zhou. 2010. Methods of analysis for GNSS signal quality (in Chinese). Scientia Sinica Physics Mechanics and Astronomy 40(5): 528–533.

Stanton, B.J., and R. Strother. 2007. Analysis of GPS Monitor Station Outages. Pp. 176–183 in Proceedings of the 20th International Technical Meeting of the Satellite Division of the Institute of Navigation (ION GNSS 2007), Fort Worth, Texas, September 2007.

Yang, Y.X. 2010. Progress, contribution and challenges of Compass/BeiDou satellite navigation system (in Chinese). Acta Geodaetica et Cartographica Sinica 39(1): 1–6.

Alternative Position, Navigation, and Timing: The Need for Robust Radionavigation

MITCHELL J. NARINS and LEO V. ELDREDGE
U.S. Federal Aviation Administration

PER ENGE and SHERMAN C. LO
Stanford University

MICHAEL J. HARRISON and RANDY KENAGY
Aviation Management Associates

ABSTRACT

Positioning, navigation, and timing (PNT) services are the key basis for the provision of both essential (safety and security) and economically beneficial applications worldwide in the 21st century. Whether users are ground-based or sea-based or in the air, their primary/go to source of PNT is a Global Navigation Satellite System (GNSS). While the transition of various users/modes of transport from legacy PNT aids to GNSS is at varying stages, it is of concern that the ability of users to revert from the highly accurate positioning, area navigation (RNAV), and precise time and time difference provided by GNSS back to previous methods, which may provide lower levels of performance, will require higher levels of user skills, knowledge, and abilities—capabilities that may no longer be available when needed without significant investment in equipment sustainment and upgrade and in-depth training and practice.

It is most necessary that the transition from GNSS-provided PNT services to an alternate means of achieving PNT ensures safety and security, precludes significant loss of economic benefits, and requires little change in the way operations are carried out and that it is a robust PNT solution through the development and integration of an Alternative PNT (APNT) capability. The Federal Aviation Administration (FAA) is initiating an APNT program to research various alternative strategies that will ensure that the PNT services necessary to safely, securely, and effectively support the U.S. National Airspace System's (NAS's) transition to the Next Generation Air Transportation System (NextGen) are available. This paper discusses the scope of the problem, including the extent of known and predictable and unknown and unpredictable jamming, and each of the alternative strategies identified so far, and their pros and cons.

INTRODUCTION

To properly address the need for *robust radionavigation*, it is prudent to first agree upon what is meant by the term robust. After exploring a number of sources, the most appropriate definition found, one that applies to processes, organizations, or systems and best promotes the theme of this discussion, is *the ability to withstand or overcome adverse conditions*. This then leads us to define robust radionavigation as *the provision of PNT services that are strong, sturdy, and able to withstand or overcome adverse conditions.*

For radionavigation, the term *adverse conditions* implies situations where the accuracy, availability, integrity, or continuity of the data or information carried by a radionavigation signal is impacted so as to produce unacceptable, unsafe, or unsecure results. This occurs in the presence of interference.

Interference comes in a number of different varieties. It can be intentional or unintentional. Many, if not most, instances of radionavigation interference have been from sources that were totally unaware that they were causing a problem. Interference can be predictable or unpredictable. Some radio interference is actually planned and mitigations can be put in place to minimize, if not eliminate adverse effects. Interference can be both manmade and environmental. Recently much discussion has occurred in regards to our solar cycle and how increased sunspot activity has the potential for significant impacts to GNSS-provided services. Interference can be crude or sophisticated (sometimes referred to as jamming or spoofing), the latter being much more problematic. While losing radionavigation services is never pleasant, not knowing that the services have been lost and relying on instrumentation that is faulty can be much worse. Interference can be either widespread, affecting hundreds of square miles and thousands of feet of airspace, or localized, affecting only specific operators and operations. Finally, interference can be continuous or random. While a constant-on jammer causes problems, locating one that randomly "pops up" and stays on for short periods of time can be much more problematic, because it promotes uncertainties in users—the "should I or shouldn't I" problem. In the case of safety and security operations, the answer is inevitably "I should not," making the intermittent interferer as effective, but more deceptive than the constant interference source.

Still, when assessing whether a condition is adverse, one must do so in light of the radionavigation system being employed—both on the transmitter and receiver ends. What is adverse for one may not be adverse for another, and that is a basis for determining an appropriate alternative PNT strategy that ensures safety and security and minimizes the impact to the economy. Some PNT systems rely on extremely low-power signals while others employ high-power transmissions. Some rely on line-of-sight signals, while others employ ground waves. Some have been designed from the start to work in *adverse conditions*, while others expect every day to be a sunny day.

The message is that the world has already changed and is still changing. Interference occurs more and more often—from both predictable and unpredictable sources. The most prudent course of action by both suppliers and users of radionavigation services is to ensure that they fully appreciate the potential for real-world interference and plan and design accordingly.

GROWING SOURCES OF INTERFERENCE

Certainly the most *predictable* source of interference to GNSS-provided PNT is exercises conducted by military organizations, whose missions require them to be able to both deny services to opposing forces and operate in GNSS PNT-denied situations. To ensure their readiness, a significant amount of testing is required. Figure 1 denotes the locations, extent, and duration of GNSS interference events originating from the U.S. Department of Defense (DOD) sources. To ensure that neither the FAA nor the DOD mission is impaired, FAA and DOD coordinate these exercises to ensure that the safety, security, and economic benefits of the U.S. NAS are not impaired and that the need for DOD readiness is properly supported.

However, unpredictable interference is much more insidious and is becoming a much bigger problem day by day, driven in part by peoples' awareness that the GNSS receiver in their car or mobile phone allows others to track their location. In

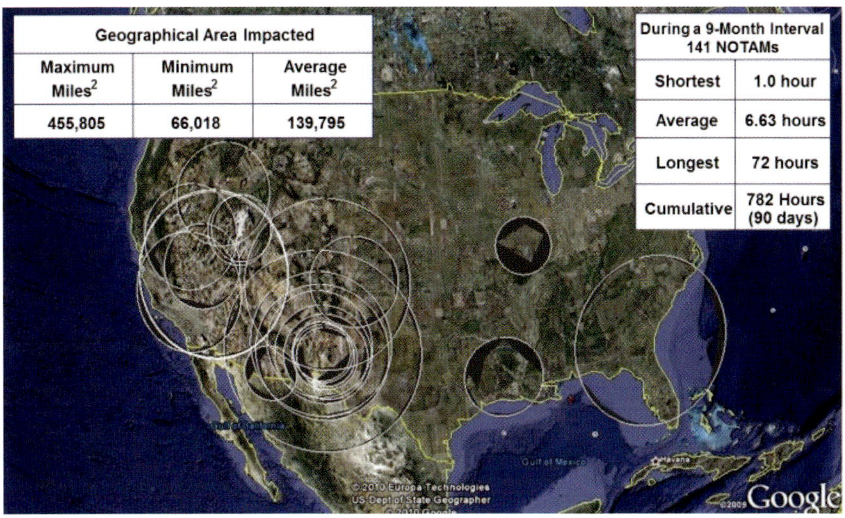

FIGURE 1 Adverse condition: GPS testing by DOD.

response, a number of manufacturers have produced what they call *personal protection devices*, small, compact jamming devices that are sold to either interfere only with GNSS signals or to jam both GNSS and cellular telephone transmissions. Figures 2, 3, and 4 provide images of just some of the devices that, while illegal in most parts of the world, are readily available on the Internet.

According to its specifications, also available on the Internet, the jamming device shown in Figure 2 is capable of transmitting 0.5 W of power on the Global

FIGURE 2 So-called "personal protection device."

FIGURE 3 A few more "personal protection devices."

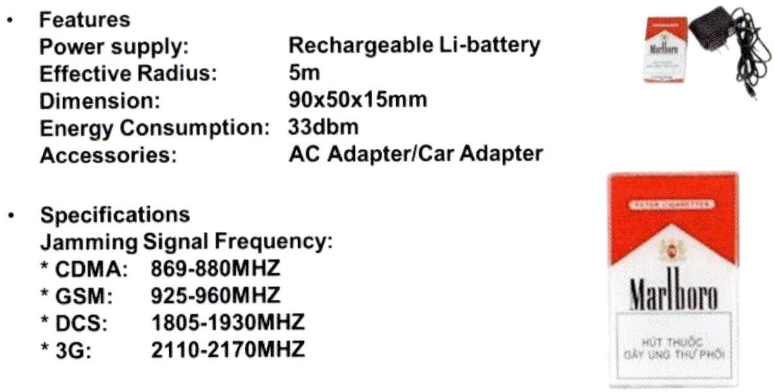

FIGURE 4 So-called "super HOT jammer cell phone jammer."

Positioning System (GPS) L1 frequency (1575.42 MHz). While it claims to be effective for only 2–10 meters, in actuality its range can extend hundreds of meters and cause significant disruption to other GNSS users—even those involved in providing safety and security services. Its price on the Internet is listed as $33.

For a bit more, personal protection devices are available that will jam multiple GNSS and cellular telephone frequencies. Some of these jammers can produce an interference signal that exceeds 5 W.

A recent addition to the jammers available on the Internet is shown in Figure 4. While it does not profess to operate on GNSS frequencies, the ability of this device to do so given the frequency ranges for which it does operate is clear. One can only imagine the effect of these devices if carried aboard planes or trains or ships or buses.

As a provider of safety and security radionavigation services that provide significant economic benefit, the FAA is keenly aware of this ever-emerging problem. That is the first step—to be aware that as a GNSS service user or supplier you are operating in harm's way. Figure 5 denotes an excellent example of this. Here, the FAA has installed a Local Area Augmentation System (LAAS), the Ground-Based Augmentation System (GBAS), at Newark Liberty International Airport (EWR), an airport that is ringed by major highways. The system's extremely sensitive GNSS antennae are located close to the New Jersey Turnpike, where literally many thousands of trucks and automobiles pass by each day—a location dictated by siting criteria based on runway configuration. Being aware of the potential problems, the LAAS program has successfully implemented system design aspects to mitigate the effects of interference sources and maintained safe and secure services. It has been a valuable lesson—one that it is hoped will be taken up by PNT users and suppliers worldwide.

FIGURE 5 In harm's way—FAA GBAS installation at EWR.

Alternative Position Navigation and Time

The FAA, in compliance with U.S. national policy, needs to maintain aviation operations indefinitely in the event of a GPS interference event or outage. This means both maintaining safety and security while minimizing any economic impact. From the FAA's perspective, a key aspect of any alternative is that NAS services can be continued throughout an interference event. Waiting for the source of the interference to be located and turned off is not an acceptable alternative.

As the FAA migrates today's NAS to the NextGen, the reliance on GNSS-provided PNT services will only grow. As NextGen evolves from a ground-based system of air traffic control to a satellite-based system of air traffic management it will rely more and more on aviation-specific GNSS-technology applications. These applications will allow more aircraft to safely fly closer together on more direct routes, thus reducing delays and providing unprecedented benefits for the environment and the economy.

To maintain safety and security and minimize impact to the economy, an alternative means of providing position, navigation, and timing services must be sought. The FAA has, therefore, initiated an APNT program to research various alternative strategies that will ensure that the PNT services necessary to safely, securely, and effectively support today's NAS and its transition to the NextGen will be ensured. An important realization is that today's air traffic control system cannot simply be scaled up to handle the predicted 2X traffic in the future. Nor can air traffic controllers handle such an increase using radar vectors. Automation and surveillance systems requiring PNT services will need to separate aircraft performing trajectory-based operations (TBOs) based on area navigation (RNAV) and required navigation performance-based (RNP) routes. Controllers will need to intercede only to provide "control by exception."

The value of RNAV/RNP is shown in Figures 6 and 7. Figure 6 shows the number of aircraft that can safely "fit" into a 10-nautical mile (nm) airspace depending on the navigation performance available. The navigation performance is a combination of the navigation service provided, the navigation capability of the aircraft avionics, and the ability of the pilot and onboard systems to fly the intended path. As you can see, the number of aircraft capable of safely using the airspace increases dramatically as the capability reaches RNP 0.3. The reason for this increase is further denoted in Figure 7.

A radionavigation/avionics system providing only RNP 1.0 capability would not be sufficient to allow aircraft to safely maintain a three-mile separation standard—the standard desired to support better airspace utilization and advanced procedures under NextGen. With RNP 0.3 capability, not only can three-mile separation be safely achieved, but also it should support procedures for parallel runway operations. It is, therefore, most important that the PNT services that support the safe, secure, and efficient operation of the NAS not be impaired and that an APNT system be developed.

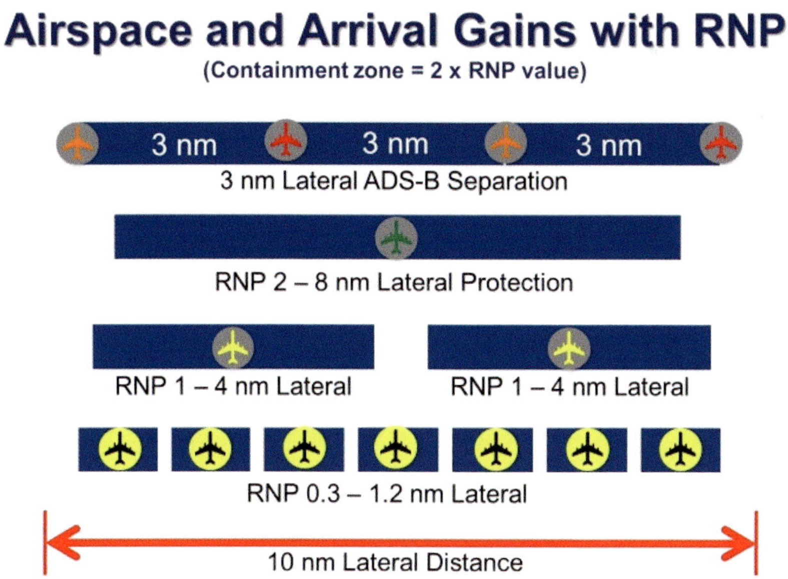

FIGURE 6 The value of RNP to airspace capacity.

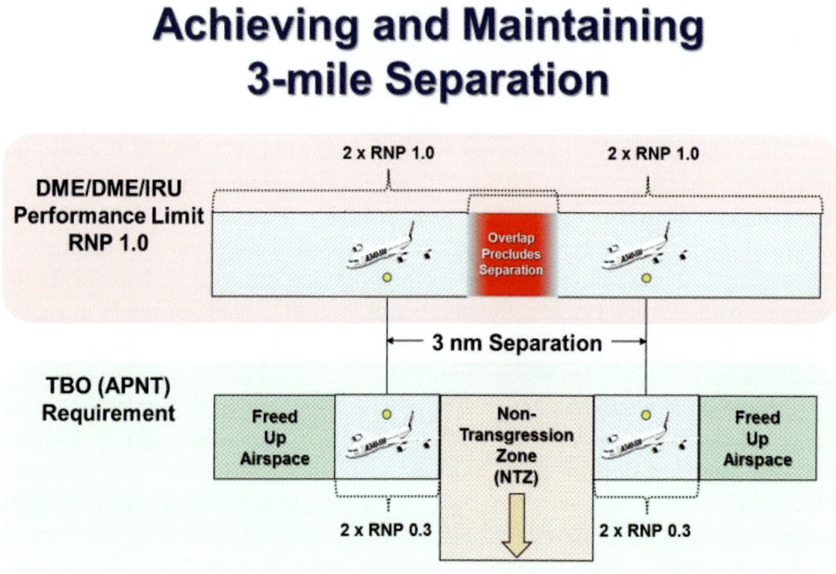

FIGURE 7 The benefit of providing RNP 0.3.

TRADE-OFFS

The determination of the solution to any problem starts with a description of the problem and a realization that trade-offs will be necessary to reach a realistic and implementable outcome. The problem statement is fairly simple—the NAS operations now and in the future will rely heavily on PNT, most PNT today and more in the future will be derived from GNSS, and GNSS-provided PNT services are vulnerable to *adverse conditions*. Figure 8 denotes the possible trade-space of solutions.

On the left are the operational contingencies that rely on procedural air traffic control. These alternatives cannot support the "normal" capacity of the NAS, so many aircraft will not be able to fly their intended routes—or in many cases, fly at all. Safety and security will be maintained, but economic impact will be great. On the right are redundant capabilities, which provide all aspects of the systems—in the air and on the ground PNT services equivalent to that provided by GNSS. Safety, security, and economic benefit are maintained for these alternatives, but the costs and resources associated with their implementation may not be realistic—especially in an industry where the potential for avionics and infrastructure changes are measured in decades rather than years. A prudent middle ground are alternatives that provide a backup capability. While not totally

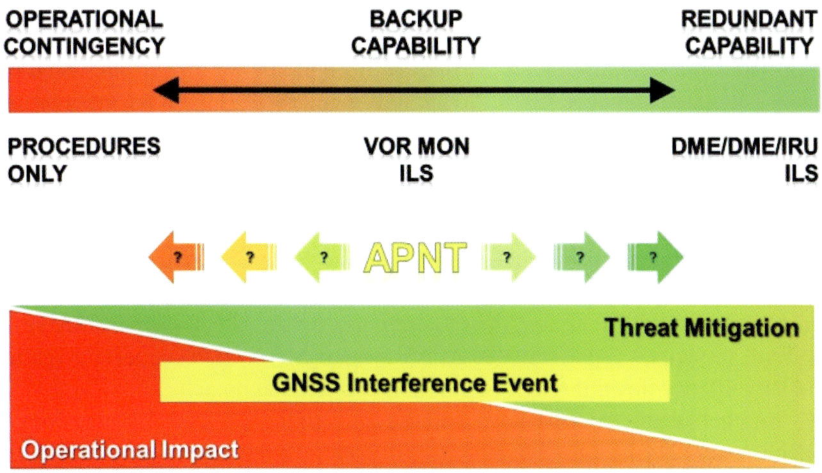

FIGURE 8 APNT trade-space.

eliminating potential economic impact, they minimize the impact to an acceptable level while ensuring safety and security are maintained.

Therefore, the goal of the FAA's APNT research is to provide a cost-effective alternative PNT service that:

- Ensures continuity of operations in NextGen
- Provides performance-based navigation (PBN)—RNAV/RNP
- Supports dependent surveillance operations (automatic dependent surveillance—broadcast, both out and in)
- Supports trajectory-based operations (TBO) and four-dimensional trajectories (4DT)
- Supports all users (general aviation, business, regional, air carrier, military)
- Minimizes impact on user avionics equipage by leveraging existing or planned equipage as much as possible
- Supports backward compatibility for legacy users
- Minimizes the need for multiple avionics updates for users
- Provides long lead transition time (circa 2020 transition)

It is also important to the FAA to avoid the potential $1.0B costs of having to recapitalize the existing very-high-frequency radio range (VOR) system that currently supplies a non-GNSS backup position and navigation capability, albeit not to the accuracy of GNSS and without area navigation capability. The VOR backup cannot support RNAV/RNP and does not provide a GNSS-independent timing capability. The FAA hopes to disestablish all VORs by 2025.

In order to determine the viability of alternative solutions, the FAA first assessed the minimum PNT requirements an acceptable alternative would need to provide. These requirements are shown in Figure 9. On the left are listed the various airspace domains, i.e., en route, terminal, LNAV (lateral navigation/non-precision approach), LPV (localizer with precision vertical), and GBAS-enabled Cat I and Cat III landings. On the right are the systems that provide the necessary capabilities to support these operations. In the middle are the navigation and surveillance requirements required for each operation—navigation measured in accuracy and containment with integrity and surveillance measured by Navigation Accuracy Category (NAC) and Navigation Integrity Category (NIC). After much analysis and discussion, the requirements for an APNT system were set at the level shown, i.e., an acceptable APNT system will need to support navigation and surveillance down through LNAV/non-precision approach.

Another key metric/parameter not discussed so far is "where." *Where* does an APNT system need to provide *what* performance? It is apparent that neither the U.S. NAS nor any other NAS is homogenous. There are key areas where capacity requirements significantly increase. In the United States, the FAA has identified 135 terminal areas where significant capacity is required and where loss of capacity would cause significant economic impact. Figure 10 denotes these areas as seen from Flight Level (FL) 180 (18,000 feet).

The FAA has categorized the airspace into three zones. Zone 1 is the airspace at FL 180 and above—all the way to FL 600 (60,000 feet). Zone 2 is the airspace that is below FL 180 and above 5,000 feet above mean sea level (MSL). Zone 3 is the

		Navigation (≥ 99.0% Availability)		Surveillance (≥99.9% Availability)			Positioning
		Accuracy (95%)	Containment (10^{-7})	Separation	NACp (95%)	NIC (10^{-7})	GNSS PNT (99.0 – 99.999%)
APNT	En Route	*10 nm	20 nm	5 nm	308m (7)	1 nm (5)	GPS
		*4 nm	8 nm				
		*2 nm	4 nm				
	Terminal	*1 nm	2 nm	3 nm	171m (8)	0.6 nm (6)	DME Only Gap
	LNAV	*0.3 nm	0.6 nm				
	RNP (AR)	*0.1 nm	**0.1 nm	2.5 nm DPA	171m (8)	0.2 nm (7)	SBAS
	LPV	16m/4m	40m/50m	2.5 nm DPA	171m (8)	0.2 nm (7)	
	LPV-200	16m/4m	40m/35m				
	GLS Cat-I	16m/4m	40m/10m	2.0 nm IPA	121 m (8)	0.2 nm (7)	GBAS
	GLS Cat-III	16m/2m	40m/10m				

* Operational requirements are defined for total system accuracy, which is dominated by fight technical error. Position accuracy for these operations is negligible.
** Containment for RNP AR is specified as a total system requirement; value representative of current approvals.

FIGURE 9 Performance-based navigation and surveillance requirements.

airspace that supports terminal operation in high-density areas. It is defined as starting 500 feet above and out to 5 statute miles (sm) from the airport, and then going up at a 2 degree angle to 5,000 feet. Figure 11 shows these three different zones.

Definition of these zones and the PNT requirements within these zones was necessary to be able to appropriately bound solutions that rely on ground-

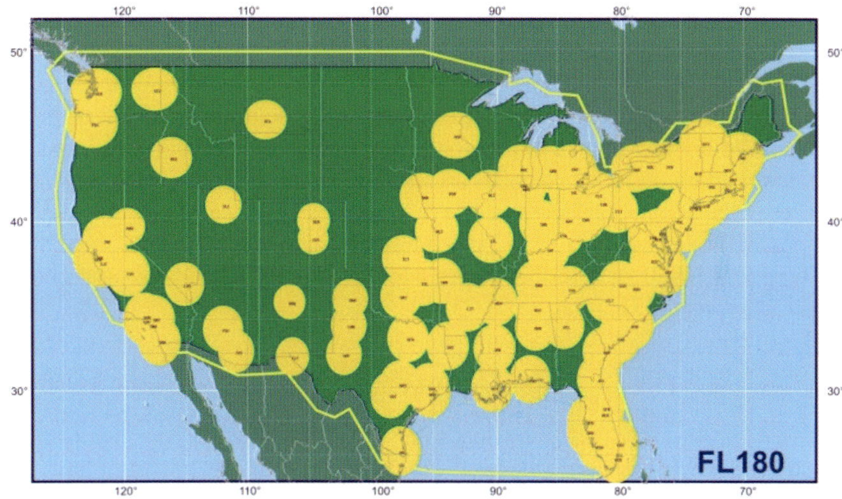

FIGURE 10 High capacity need areas in the conterminous United States (CONUS).

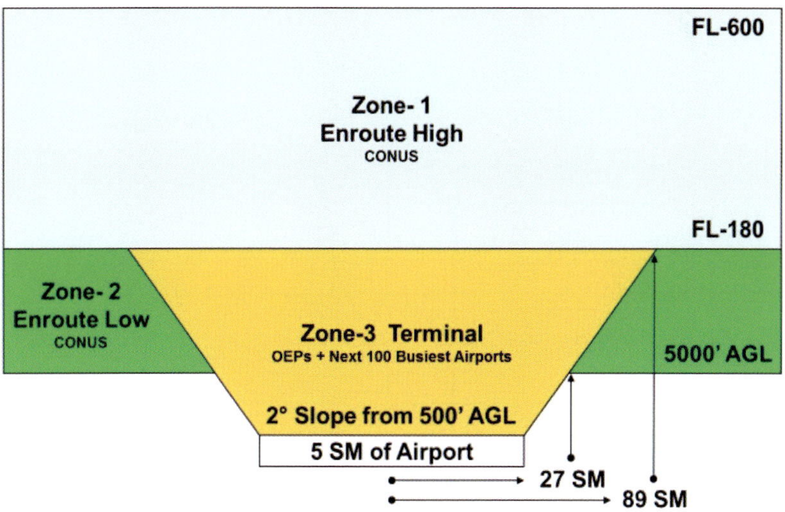

FIGURE 11 PNT performance zones.

based and line-of-sight assets. Throughout the FAA's analysis of alternatives and selection of solution(s), safety and security will always be ensured and services provided where economics warrant.

In looking for potential solutions, the FAA has concentrated on the availability of systems onboard aircraft and how to leverage existing and future equipage to facilitate an acceptable solution with a reasonable transition time. Figure 12 shows the various systems on the aircraft and where APNT solution(s) might best fit in.

APNT ALTERNATIVES

The FAA has concentrated on three categories of solutions that appear promising, while inviting input from the public and industry at meetings, symposia, and conferences on other potential areas of research. The three categories that are currently being considered are (1) Optimized Distance Measuring Equipment (DME) Network, (2) Wide-Area Multilateration, and (3) DME Pseudolite Network. Each will be described below, along with the pros and cons associated with each potential solution.

Optimized DME Network

Historically DMEs provide pilots with slant range distance from their aircraft to the DME. DMEs that are collocated with VORs traditionally provide pilots with their slant range distance to the end of an airway, while DMEs that are co-located with landing systems at airports provide pilots with their slant range to runway ends. Avionics engineers recognized that because aircraft at altitude could see a number of DMEs, a system using multiple DME ranging sources could provide

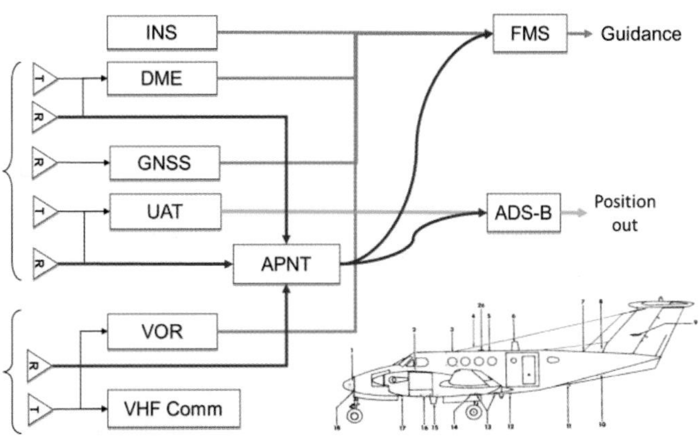

FIGURE 12 Potential APNT solutions on aircraft.

pilots with their position. However, because the DME network was not designed or laid out for this function, gaps in service coverage exist—caused by lack of DMEs or lack of necessary geometry between available DMEs. The current population of DMEs in CONUS (continental United States) is shown in Figure 13, many of which are associated with military tactical navigation (TACAN) facilities. DMEs provide high-power transmissions, typically 1,000 W.

While a DME network solution leverages existing technology and systems and will have the least impact on avionics for air carriers, there will be a significant impact on general aviation, where avionics are not available. While the FAA is planning to fill gaps in the DME coverage at FL 180 and above, this assumes that aircraft are equipped with inertial reference units (IRUs) that allow them to coast through gaps in coverage. Aircraft without IRUs are currently not authorized to fly RNAV/RNP routes and even those aircraft with an IRU are not authorized to conduct a published approach procedure requiring less than RNAV/RNP-1.0. There is also a concern that a significant increase in use of the DME network could cause interrogation saturations and impact service delivery. Finally, unless general aviation can be equipped with DME RNAV capability, there may be a need to retain and recapitalize a large number of the VORs at a substantial cost.

Wide Area Multilateration

Wide Area Multilateration (WAM) utilizes signals that are transmitted frequently from an aircraft equipped with ADS-B to determine the aircraft's position. Figure 14 denotes the sequence of events that occur that would allow an aircraft to learn its position in the event of a loss of GNSS-provided PNT.

FIGURE 13 1,100 DMEs in CONUS.

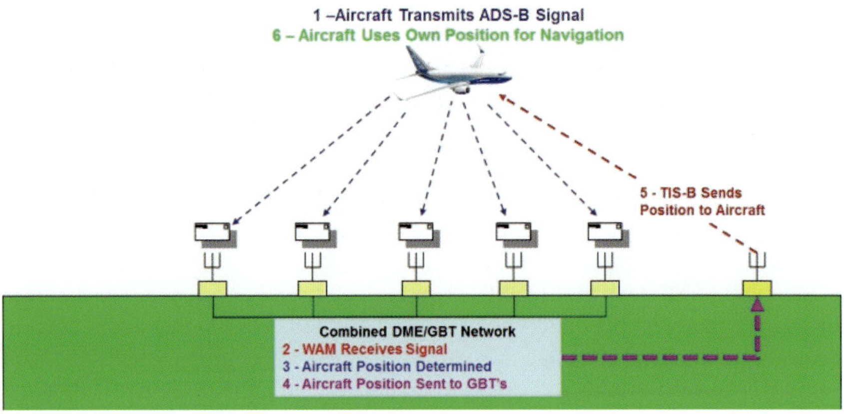

FIGURE 14 Passive wide area multilateration.

Ground-based transceivers (GBTs) being installed to support ADS-B can utilize this technology to determine aircraft position in the event that the aircraft cannot. The national ADS-B GBT system is shown in Figure 15.

By leveraging the DME installed base and the planned GBT installations, coverage across CONUS would be greatly improved. Figure 16 shows this combined infrastructure.

While the WAM solution will have a minimal impact on existing avionics for surveillance, integrity monitoring and Time-to-Alert necessary to meet navigation requirements may be very challenging. Still, accuracy has been demonstrated to be within target levels and it is compatible with existing WAM systems. There are, however, concerns regarding the availability of bandwidth on the 1090 MHz extended squitter channel so that capacity may be limited in high-density environments. Use of WAM for navigation will also entail changes to existing avionics.

WAM also requires that each of the ground stations maintains a common time reference as WAM is a time-of-arrival system. Current systems utilize a common beacon that can "be seen" by all systems as the synchronizing mechanism. Wider-area systems may encounter issues and certainly additional costs if beacons were the only means to maintain synchronization.

DME Pseudolite Network

DMEs broadcast in the L-band, the same area of the spectrum as GNSS. They work by receiving interrogations from aircraft and replying after a fixed delay, thus allowing the aircraft to determine its slant range to the DME. When a DME is not being interrogated it maintains a "heartbeat" awaiting the next interrogation. The DME Pseudolite (DMPL) solution would include a transmission on the DME

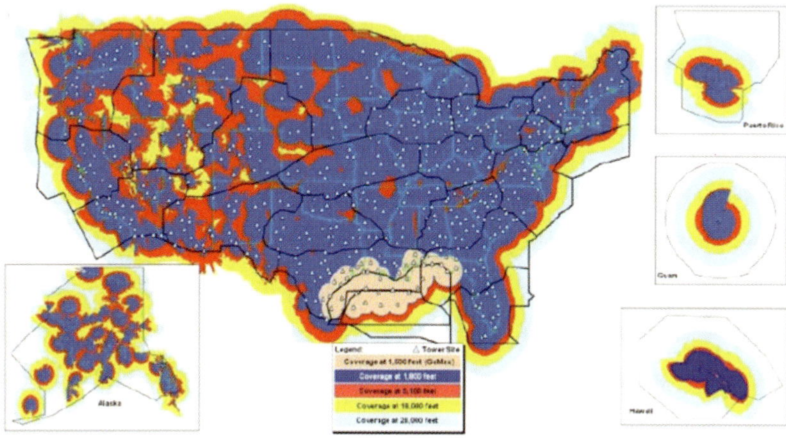

FIGURE 15 800 GBTs to be installed nationwide.

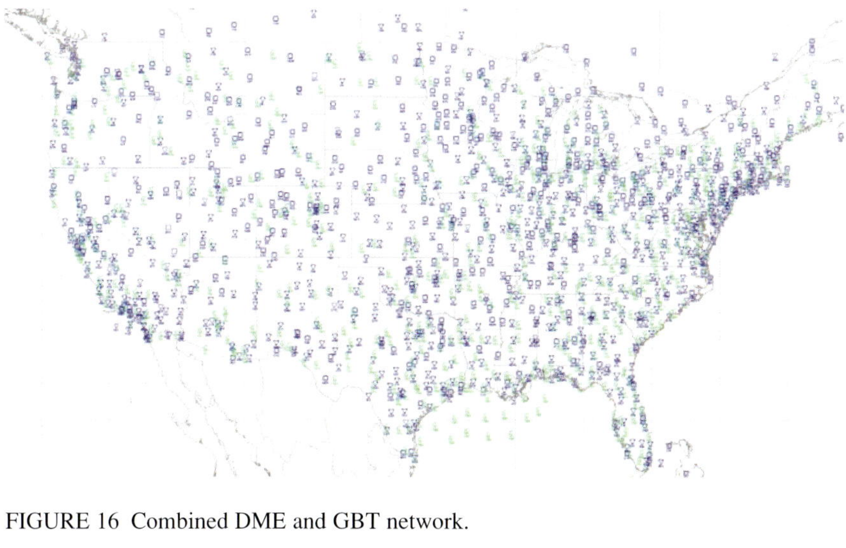

FIGURE 16 Combined DME and GBT network.

heartbeat, which would be maintained continuously, thus identifying the particular DME, its location, and the time-of-day. The aircraft, using the same methodology employed by GNSS and WAM systems, would determine its position. As the aircraft would receive the "raw" data, it would be left to the aircraft to determine the integrity of the derived information, just as it does for GNSS.

The DMPL alternative provides unlimited capacity and an aircraft-based position and integrity solution and could leverage use of existing DMEs and

GBTs. However, it would require modification to the DME transmit signal. It would require a minimum of three sites to compute aircraft position (unless the DMEs interrogation/reply capability was also utilized, and then two would suffice). The DMPL alternative would also require a common GNSS-independent timing reference similar to that needed by the WAM solution. While it would have the greatest impact to aircraft avionics, it could potentially provide the most benefit. There is the potential to include position calculation and integrity monitoring functions in ADS-B in avionics. Because it is the least mature concept, no avionics are yet in development and no standards have been established. It would also require the retention and recapitalization of nearly half the VORs unless general aviation is equipped with Pseudolite avionics.

Time Synchronization

The need to provide time synchronization for both the WAM and DMPL alternatives, as well as the need to provide frequency services for telecommunication applications, caused the FAA to research alternative time and frequency provision as part of the APNT effort. During the problem analysis phase the FAA determined that if the sources of GNSS interference were so great as to preclude use of any satellite in any direction, the situation would be outside the FAA's means to mitigate the time service interruption. Therefore, the FAA assumed that the interfering source would arrive from at most a few directions and that by utilizing a steerable null antenna, the jammer could be substantially eliminated and a source of good time and frequency reinforced. Figure 17 shows how this concept would work.

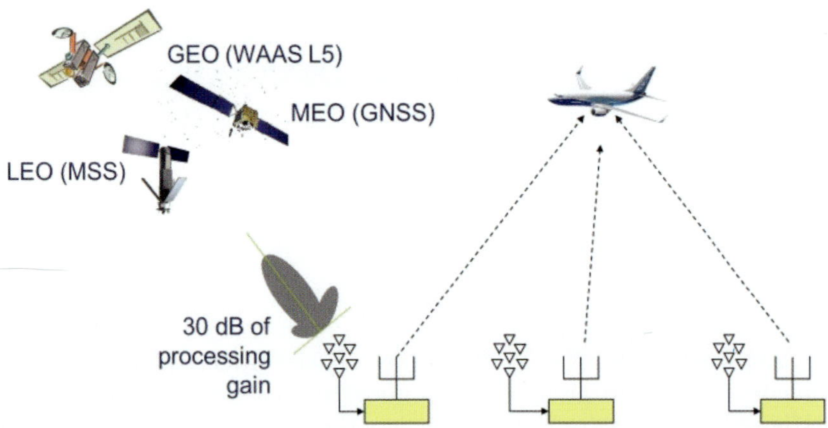

FIGURE 17 Ground-based time synchronization.

Steerable null antennas located at ground facilities (either DME or GBT) should be able to sufficiently null out interfering signals while reinforcing the time and frequency signals from a satellite—whether it be in GEO, medium, or low Earth orbit. This would allow GBTs or DMPL or both to continue providing multilateration services despite a GNSS service interruption.

NEXT STEPS

In pursuit of the best APNT solution(s), the FAA is developing a Project Plan for Full Investigation, the means to validate backup requirements and is performing appropriate system engineering analyses. The FAA plans to develop R&D prototypes along with cost schedule estimates while it completes the analysis of alternatives. The schedule for accomplishing these actions is show in Figure 18.

First and foremost, the APNT remains a research endeavor. The "best" answer is still, as they say, *to be determined*. What is most important is that the potential problems and impacts have been recognized and steps are being taken to ensure that the safety, security, and efficiency of the U.S. NAS will be maintained in the event of a loss of GNSS-provided PNT.

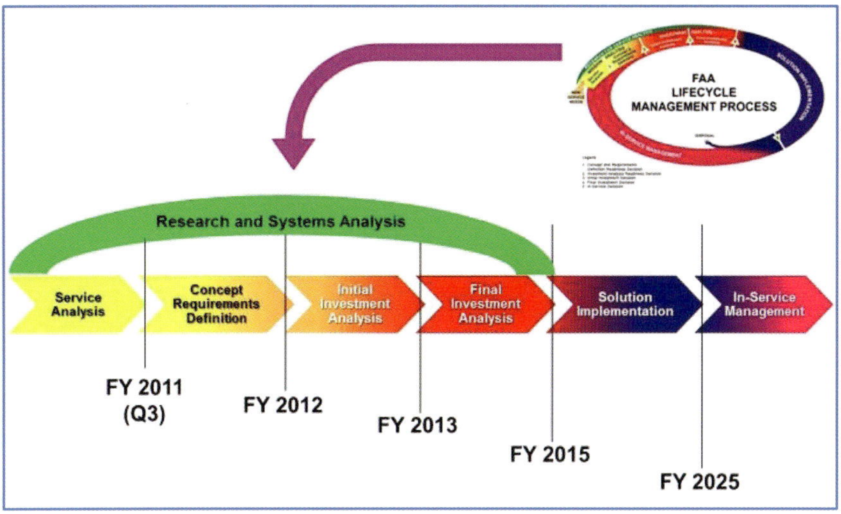

FIGURE 18 APNT program life cycle.

Analysis of the GNSS Augmentation Technology Architecture

CHEN JINPING
Beijing Global Information Center of Application and Exploitation

BACKGROUND

The development process of GNSS is divided into four stages: (1) experimental construction stage (GPS I), (2) infrastructure construction stage (GPS II), (2) augmentation construction stage (Wide Area Augmentation System [WAAS]/ Local Area Augmentation System [LAAS]), and (4) architecture construction stage (GPS III).

The architecture construction stage is to meet the requirements from different military and civilian users, consider multi-GNSS compatibility and interoperability, design the basic GNSS and augmentation system in whole, and emphasize the policy, law, standards, and industrialization of GNSS.

The current existing augmentation technologies were developed for legacy GPS, and the augmentation systems are independent of each other. There are no uniform definition and standard. The user is a little confused with the application of these augmentation systems.

In the architecture construction stage, the redefinition of GNSS augmentation technology architecture is needed. Future augmentation systems should be constructed for the improvement of basic GNSS performance.

INTRODUCTION TO THE CURRENT GNSS AUGMENTATION TECHNOLOGY

The current GNSS augmentation technologies are shown in Figure 1. The comparison of different GNSS augmentation technologies is shown in Tables 1 and 2.

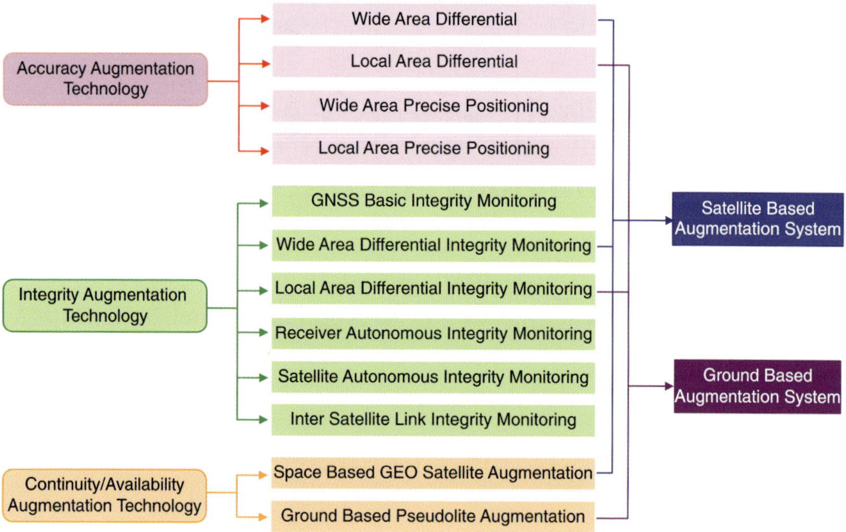

FIGURE 1 The current GNSS augmentation technologies.

ANALYSIS OF GNSS AUGMENTATION GOALS AND OBJECTS

The goal of GNSS augmentation is to meet the improved performance requirements of the high-level user. GNSS navigation performance requirements defined by the International Civil Aviation Organization (ICAO) and the Radio Technical Commission for Aeronautics (RTCA) are shown in Tables 3 and 4.

The above tables show that the accuracy, integrity, continuity, and availability requirements are different in different flight phases. The accuracy requirement can be met easily, but the integrity, continuity, and availability requirements are stringent and cannot be provided by the current GNSS.

In addition, some high-level users have required decimeter and centimeter positioning accuracy, and these requirements cannot be provided by the current GNSS either.

The objects of augmentation are to allow the basic GNSS, including GPSII, modernized GPS, GLONASS, Galileo, COMPASS regional system and GPSIII, modernized GLONASS, Galileo+, and COMPASS global system, to meet positioning, navigation, and timing service requirements.

The constellation, user range error (URE) performance, and integrity monitoring performance are different for different GNSS and for their different construction stages. So the accuracy, integrity, continuity, and availability performance are at different levels (Table 5).

The goals of augmentation show the accuracy requirements of augmentation can be divided into two classes: 1 m level and <1 m level (including decimeter

TABLE 1 Comparison of Current GNSS Differential Technologies

	Coverage and Station	Processing Principle	Broadcasting	Performance	Status
Wide Area Differential Technology	Thousands of kilometers, Tens of stations	State differential method, generation of ephemeris, clock correction and ionospheric delay parameters and based on pseudorange measurements (assisted with carrier phase)	GEO/radio beacon, RTCA	Accuracy 3 m (single frequency), no initialization	WAAS EGNOS MSAS GAGAN
Wide Area Precise Positioning Technology	Global, Nearly 100 stations	State differential method, generation of ephemeris, clock correction and mainly based on carrier phase measurements	GEO/Internet, SOC or self-defined protocol	Accuracy dm level (dual frequency carrier phase), 20 minutes initialization	GDGPS StarFire OmniStar
Local Area Differential Technology	Tens of kilometers, Several stations	Measurement differential method, generation of integrated pseudorange correction parameters and mainly based on pseudorange measurements (assisted with carrier phase)	VHF, RTCA/RTCM protocol	Accuracy <1 m (single frequency), no initialization	LAAS NDGPS RBN-DGPS
Local Area Precise Positioning Technology	Tens of kilometers, Several stations, or extended to thousands of stations	Measurement differential method, mainly based on carrier phase measurements, Local area users can receive the local area error correction parameters processed with the VRS, FKP, MAC modes	GPRS/CDMA, RTCM protocol	Accuracy cm level (dual frequency carrier phase), 1~2 minutes initialization	CORS

TABLE 2 Comparison of Current GNSS Integrity Monitoring Technologies

	Coverage and Station	Processing Principle	Broadcasting	Performance	Status
GNSS Basic Integrity Technology	Global, Several to tens of stations	Generation of URA or SISA/SISMA with processing of ephemeris and clock error	Nav Message	GPS, Risk, 1E-4/hr TTA, hrs to mins, Galileo, TTA, 6s Risk, 2E-7/150s	GPS Galileo
WAAS Integrity Technology	Thousands of kilometers, Tens of stations	Generation of UDRE, GIVE with processing of ephemeris, clock error and ionospheric grid parameters	GEO RTCA	TTA, 6s Risk, 2E-7/150s	WAAS EGNOS MSAS GAGAN
LAAS Integrity Technology	Tens of kilometers, Several stations	Generation of integrity parameters with processing of differential pseudorange integrated correction parameters	VHF RTCA/ RTCM	TTA, 2s Risk, 2E-9/15s	LAAS NDGPS RBN-DGPS
RAIM Technology	Visible satellites measurements	Detection and elimination of fault satellite measurements with multi measurements redundancy	—	Almost no TTA Pmd, 1E-3	Receiver
SAIM Technology	Satellite feedback signal measurements	Generation of integrity parameters with monitoring of satellite signal power abnormality, pseudo code abnormality, clock error overrun and navigation data mistake	Nav Message	TTA, 2s Risk, 1E-7/hr	Under research
ISL Integrity Technology	Inter satellite crosslink measurements	Generation of integrity parameters with monitoring satellite orbit and clock error abnormality	Nav Message	Unspecific	Under research

TABLE 3 GNSS Navigation Performance Requirements Defined by ICAO

Flight Phase	Accuracy (95%)		Integrity				Continuity Risk	Availability
	H	V	HAL		TTA	Risk		
			H	V				
Ocean	3.7 km	N/A	7.4 km	N/A	5 min	1×10^{-7}/h	$1 \times 10^{-4} \sim 1 \times 10^{-8}$/h	0.99~0.99999
Domestic	3.7 km	N/A	3.7 km	N/A	5 min	1×10^{-7}/h	$1 \times 10^{-4} \sim 1 \times 10^{-8}$/h	0.99~0.99999
Terminal	0.74 km	N/A	1.85 km	N/A	15 s	1×10^{-7}/h	$1 \times 10^{-4} \sim 1 \times 10^{-8}$/h	0.99~0.99999
NPA	220 m	N/A	556 m	N/A	10 s	1×10^{-7}/h	$1 \times 10^{-4} \sim 1 \times 10^{-8}$/h	0.99~0.99999
APV I	16 m	20 m	40 m	50 m	10 s	2×10^{-7}/P	8×10^{-6}/15 s	0.99~0.99999
APV II	16 m	8 m	40 m	20 m	6 s	2×10^{-7}/P	8×10^{-6}/15 s	0.99~0.99999
Cat I	16 m	6~4 m	40 m	15~10 m	6 s	2×10^{-7}/P	8×10^{-6}/15 s	0.99~0.99999

TABLE 4 GNSS Navigation Performance Requirements Defined by RTCA

Flight Phase	Accuracy (95%)		Integrity				Continuity Risk	Availability
	H	V	HAL		TTA	Risk		
			H	V				
Cat I	16 m	4 m	40 m	10 m	6 s	2×10^{-7}/150 s	8×10^{-6}/15 s	0.99~0.99999
Cat II	5 m	2.9 m	17 m	10 m	2 s	1×10^{-9}/15 s	4×10^{-6}/15 s	0.99~0.99999
Cat IIIA	5 m	2.9 m	17 m	10 m	2 s	1×10^{-9}/15 s	4×10^{-6}/15 s	0.99~0.99999
Cat IIIB	5 m	2.9 m	17 m	10 m	2 s	1×10^{-9}/15 s(V) 1×10^{-9}/30 s(H)	2×10^{-6}/15 s/15 s(V) 2×10^{-6}/30 s(H)	0.99~0.99999

TABLE 5 Analysis of Different GNSS Performance

		Constellation	Accuracy	Integrity	Notes
GPS	Legacy GPSII	24 satellites	URE 8 m Position >10 m	Weak	
	Modernized GPS II	30 satellites	URE ~1 m Position <10 m	Improved, but unspecific	
	GPS III	30 satellites	URE <1 m Position ~1 m	Cat I	
GLONASS	Legacy GLONASS	24 satellites	Comparative to Legacy GPSII		
	Modernized GLONASS	24 satellites	Comparative to Modernized GPSII		
Galileo		30 satellites	URE <1 m Position 3~5 m	Global SOL service (Cat I) MEO broadcast I/Nav TTA: 6 s Risk: 2E-7/150 s	Galileo+ More improvement
COMPASS	Regional System	12 satellites	URE ~2 m Position ~10 m	Augmentation system is integrated in the basic GNSS TTA:6 s Risk: 2E-7/approach (Cat I)	
	Global System	30 satellites	Comparative to Galileo and GPSIII		

and centimeter). The integrity requirements of augmentation can be divided into two classes: Cat I level (TTA 6 s, Risk 1E-7/approach) and better than Cat I level (TTA 2 s, Risk 1E-9/15 s). The continuity and availability requirements of augmentation are corresponding to the integrity requirements. Additionally, the 1 m level users are high in real time, and their integrity, continuity, and availability requirements are also high. The <1 m level users are slow in real time, and their integrity, continuity, and availability requirements are relatively low.

The objects of augmentation show: In earlier stages, for legacy GPS and GLONASS, the constellations consist of 24 satellites, the position accuracy is >10 m, and the integrity performance is weak. So the augmentation technologies and augmentation system is needed to improve the performance. In modernization stage, for modernized GPSII, Galileo, modernized GLONASS, etc., the constellations consist of 30 satellites, the position accuracy is <10 m and approaching to 1 m level step by step, the integrity performance has been improved and approaches CAT I, but the performance goals of these systems are unspecific. In architecture stage, for GPSIII, Galileo+ and COMPASS global system, the constellations consist of 30 satellites, these systems are interoperable, the position accuracy is 1 m level, and the integrity performance reaches CAT I performance. The augmentation systems should be redesigned carefully.

DEFINITION OF FUTURE GNSS AUGMENTATION TECHNOLOGY ARCHITECTURE

The current existing augmentation technologies are presented for the earlier GNSS. The future augmentation technology architecture can be defined with the following principles:

1. The 1 m accuracy and Cat I integrity performance will be provided by the basic GNSS as the first layer in global coverage and by the wide area augmentation system, which is constructed by some country or organization as second layer in regional coverage.
2. The <1 m accuracy performance will be provided by the local area precise positioning system constructed by each country or organization.
3. Better than Cat I level integrity performance will be provided by the local area integrity augmentation system constructed by a specific user group.
4. The continuity and availability performance will be assured by the fact that every constellation consists of 30 satellites and constellation interoperability.

The definition of future GNSS augmentation technology architecture is shown in Figure 2.

GNSS layer: 1 m accuracy performance is provided by the constellation and signal-in-space (SIS) user range of multi-GNSS. Cat I integrity performance

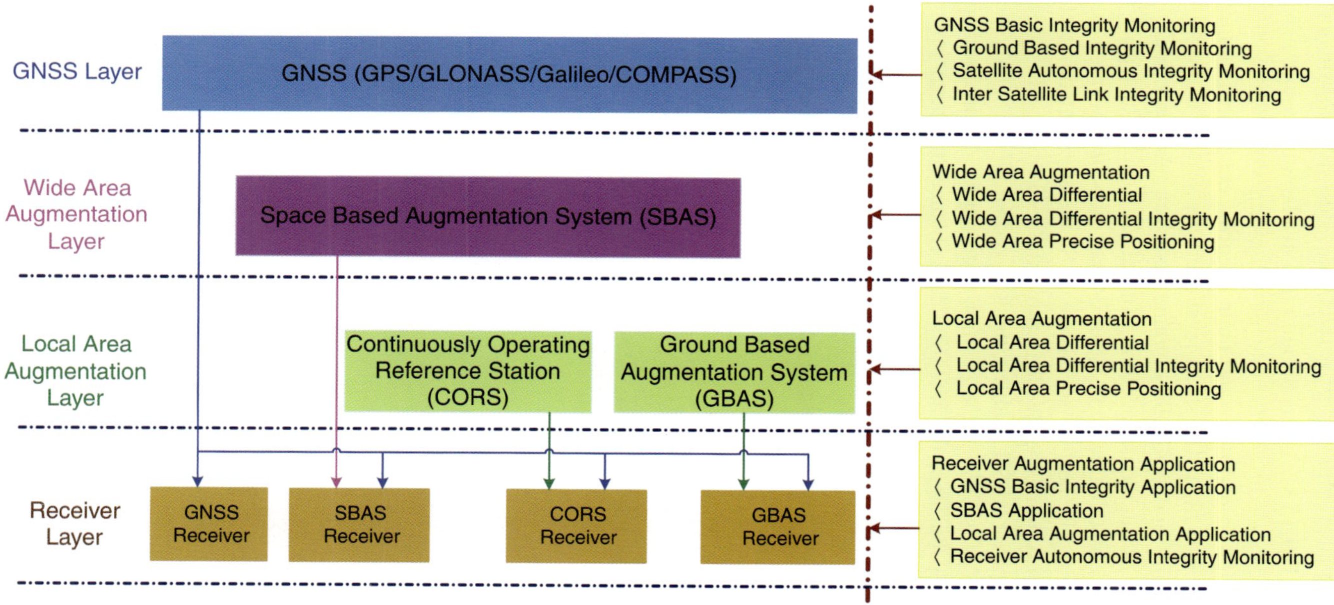

FIGURE 2 Definition of future GNSS augmentation technology architecture.

is provided by integrating the GNSS basic integrity monitoring and satellite autonomous integrity monitoring (SAIM) and inter-satellite link (ISL) integrity monitoring as the first layer in global coverage.

Wide area augmentation layer: Based on the basic GNSS, as the second layer in corresponding areas, the performance of 1 m accuracy and Cat I integrity are provided by wide area argumentation systems, which are constructed by various countries or organizations by integrating wide area difference, wide area precise positioning, and integrity monitoring technology.

Local area augmentation layer: Based on the basic GNSS, <1 m accuracy performance is provided by local area precise positioning systems, which are constructed by various countries or organizations with local area precise positioning technology. Better than Cat I integrity performance is provided by local area integrity argumentation systems, which are constructed by various user groups with local area integrity monitoring technology.

Receiver layer: Corresponding to the above three level services, there are four types of receiver for navigation, positioning, and integrity monitoring. At the same time, receiver autonomous integrity monitoring technology is used for related integrity analyzing.

CONCLUSIONS AND QUESTIONS

From the definition of the technology architecture, the GNSS basic system and the augmentation system are independent as well as coupled and reflect the multiplayer architecture. The development of GNSS basic system and augmentation system may refer to this architecture, so as to realize interoperability step by step in the aspects of design, technique realization, application standard, etc.

Of course, multi-GNSS basic system and augmentation system have already formed their definition and been constructed. To implement this architecture, the following questions should be answered:

1, For the performance of 1 m accuracy and Cat I integrity, the following questions should be studied: How to define the corresponding performance standard, such as the constellation geometry and the SIS URE? How to integrate the technologies of system basic integrity monitoring and SAIM and ISL integrity monitoring?
2. Based on the performance of GNSS basic systems, the following questions should be studied: How to classify definition for GNSS basic systems and the Wide Area Augmentation Systems constructed by different countries or organizations? How to construct independently under the same standard? How to integrate Wide Area Augmentation System and Wide Area Precise Positioning System?
3. Today, many countries are developing local precision positioning systems and local integrity augmentation systems for the legacy GPS. The

following questions should be studied: How to adapt the development and change of GPS signal? How to design and construct for multi-GNSS?

4. For different service mode and different user receiver, the following questions should be studied: How to define the standards of navigation and integrity application processing? How to define the relation between system integrity processing and receiver autonomous integrity monitoring?

Impact of Intentional, Low Power, In-Band, Personal Privacy Devices (PPDs) on Aviation

A.J. VAN DIERENDONCK
AJ Systems
Los Altos, California

What are PPDs? PPDs are small "jammers" that plug into cigarette lighter sockets that are meant to "jam" or "mask" Global Positioning System (GPS)-Global Navigation Satellite Systems (GNSS) receiver/navigators that report, through other means, the position of the host vehicle, when the user does not want his position known. Use is illegal in the United States. They are probably purchased via the Internet. Unknown (or known) to the owner, PPDs can jam other users of GPS (GNSS), including Safety-of-Life (SOL) operations. At Newark, New Jersey, airport (EWR), such a device's radio frequency interference (RFI) affected the Ground-Based Augmentation System (GBAS) operation. GBAS performance interruptions have been observed since November 2009, impacting GBAS service availability, although there was no evidence of direct effects on airborne avionics. It was determined that RFI emanated from sources on the New Jersey Turnpike (NJT) (in motion). A test program determined direction of travel, average velocity, how frequently emitter(s) traveled past the EWR GBAS, and what time of day and days of the week events were observed.

RFI was also detected at Wide Area Augmentation System (WAAS) Reference Stations (WRSs) near Washington, D.C., and Houston, Texas. However, because of diverse location and redundancy of WAAS reference receivers, no interruptions of WAAS operations were noticed.

RFI was characterized by analyzing data collected from GPS receivers, spectrum analyzers, and time samples captured by real time data collection (snapshots). Through measurements of spectral shape, time series demodulation, magnitude of degradation in GPS receivers, and estimates of emitter(s) transmitted power, RFI was characterized. The fact that the interference source was moving complicated the characterization.

THE SCENARIO AT EWR

NJT runs by the GBAS facilities (Figure 1). There was intermittent severe degradation or loss-of-lock in GBAS reference receivers. There were also additional GPS receivers/antennas installed at the GBAS shelter to record more detailed information when severe GBAS degradation was noted—one antenna with moderate directivity that was pointing horizontal northward along NJT, and one omni-directional pointing up. Of course, received interference power was greatest during the closest approach of the interference source. This resulted in negative elevation angles, so gain was low, but approximate azimuth direction could be observed.

The spectral peak of Antenna 1 is observed at GPS TOW1 (GPS Time of Week of event 1); the spectral peak of Antenna 2 is observed at GPS TOW2

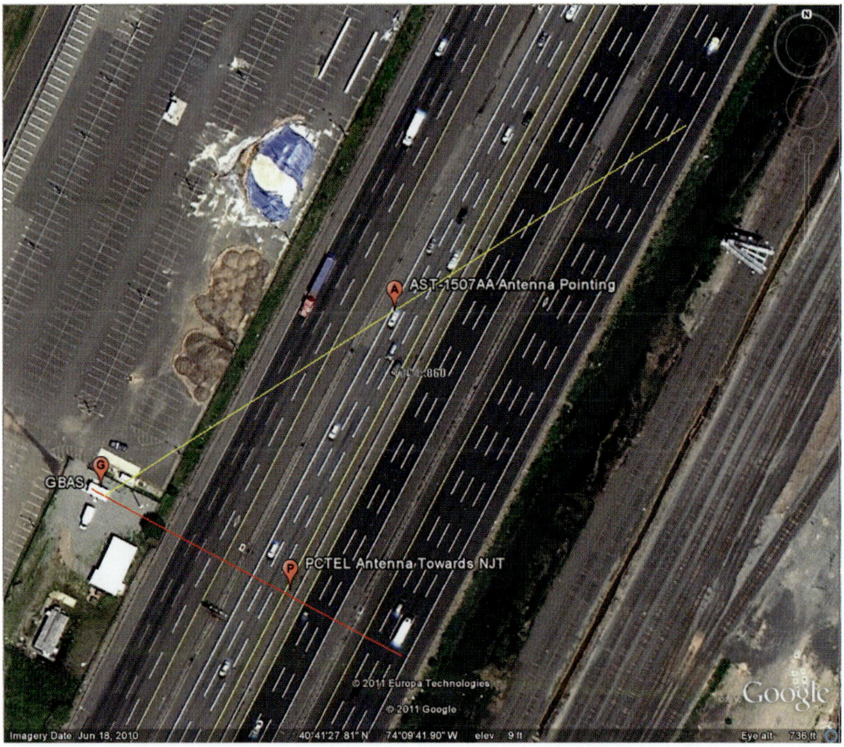

FIGURE 1 The scenario at EWR.

(event 2). The average speed is based upon the time difference. *Google Earth* images of 654 meters at 33° places RFI in the NJT southbound lane 5.

Multiple emitter types were observed (Figure 2). A linear sweep indicates that the underlying signal for this particular event is swept CW, starting at 1570 MHz and increasing to 1590 MHz in about 116 microseconds (8.6 kHz sweep rate). The small increase in the center of the band is GPS coarse/acquisition code.

OBSERVED L1 C/N$_0$ DEGRADATION—RFI POWER ESTIMATES

Figure 3 presents a histogram of Normalized Measurements of C/N$_0$ (L1 Carrier Power-to-Noise Density ratio) from a WAAS G-II Receiver that

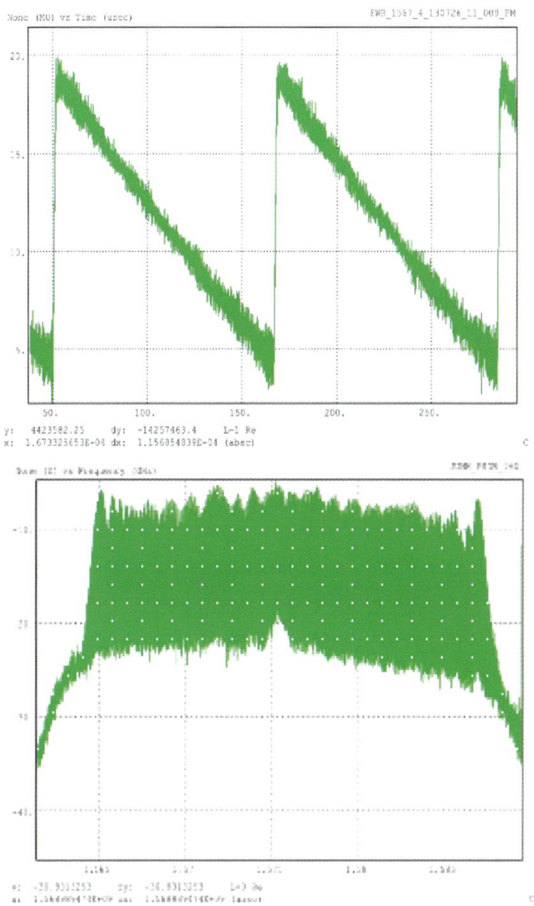

FIGURE 2 Typical wideband RFI.

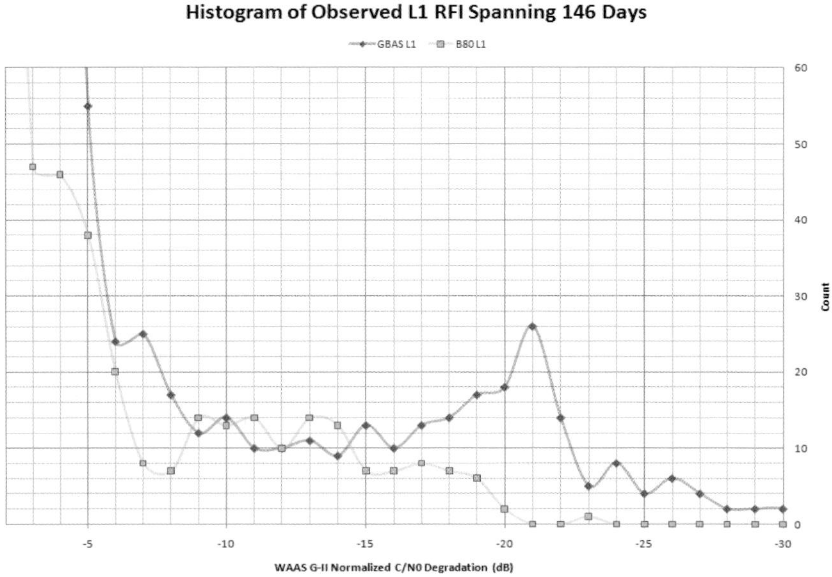

FIGURE 3 WAAS G-II normalized C/N_0 degradation (dB).

was developed covering 146 days. The events were recorded whenever loss-of-lock or severe degradation occurred in the GBAS Reference Receiver. The magnitude of degradation varied more than 20 dB, greater than the uncertainty with respect to Free-Space-Loss (FSL).

The estimated RFI power variation due to measurement accuracy is minimal—approximately ±1 dB RSS. The estimated RFI power variation due to other uncertainties is significant—transmit antenna gain and the loss exiting the vehicle, and the exact location of RFI transmitter (10 to 11 dB). The polarization mismatch loss uncertainty is 1 to 6 dB. Note that most of the degradations were less than about 8 dB, but a significant number had a degradation of 21 dB.

THE SCENARIO NEAR WAAS REFERENCE STATIONS AT LEESBURG, VIRGINIA, AND HOUSTON, TEXAS

Figure 4 shows a scenario near the WAAS Reference Station (WRS) at Leesburg, Virginia, showing highways relative to the ZDC-WRS. Figure 5 shows the degradation of Houston WAAS Reference Receiver normalized C/N_0 versus time-of-day, occurring twice each day. It is postulated that the vehicle passed by on the airport access road to the airport, and then returned later. Similar degradations were observed at the ZDC-WRS.

FIGURE 4 ZDC-WRS location relative to roads passing by.

SUMMARY

Aviation must consider PPD proliferation as inevitable. The key uncertainty in the presented results is the radiation pattern for emitter(s) installed in a vehicle. It is highly dependent on the structure of the vehicle and location of the RFI transmitter within the vehicle, which is not easily modeled. GNSS ground RF operating environment requirements may have to be improved by "hardening" from sighting and operational considerations. However, is 15–25 dB improvement even possible?

We must evaluate implications for avionics. No interruptions of GPS reception by avionics have been reported to date. Should the Avionics Minimum Operational Performance Standards (MOPS) RFI mask address unintentional *intentional* emitters, while retaining the aggregate of –110.5 dBm/Hz within +/– 10 MHz? Should we add an A/J requirement to MOPS?

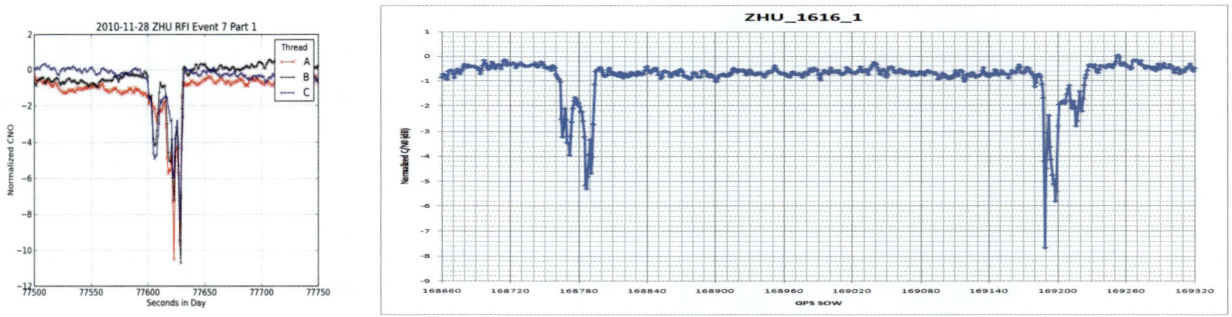

FIGURE 5 Degradation of normalized C/N_0 versus GPS time-of-day (in seconds) at Houston WRS.

GNSS Open Signals Interference Issues and Countermeasures

DU XIAODONG
Aerospace Long March Launch Vehicle Technology Co. Ltd

WANG FEIXUE and NIE JUNWEI
National University of Defense Technology

BACKGROUND

Global Navigation Satellite Systems (GNSS) can provide users with accurate PNT information services. It has already been used in various fields of national economy and peoples' lives, such as communication, traffic, electric power, finance, security, dangerous goods management, and so on. These applications bring forward a critical demand on the continuity and accuracy of GNSS.

GNSS open signals have low power when arriving at the ground, low chip rate, and short period of PN codes, so they are easily interfered with by various electromagnetic signals. Therefore, we should pay attention to the interference issues of GNSS civil signals.

BeiDou-2 is planning to provide regional regional navigation satellite services in 2012 and will face the same issues. Figure 1 shows the BD-2 system space components.

CIVIL SIGNALS INTERFERENCE TYPES AND SOURCES

Interference Types

Civil signals will face two kinds of interference—intentional interference and unintentional interference. The intentional interference consists of jamming and spoofing. Figure 2 shows the types of interference.

FIGURE 1 BD-2 system space components.

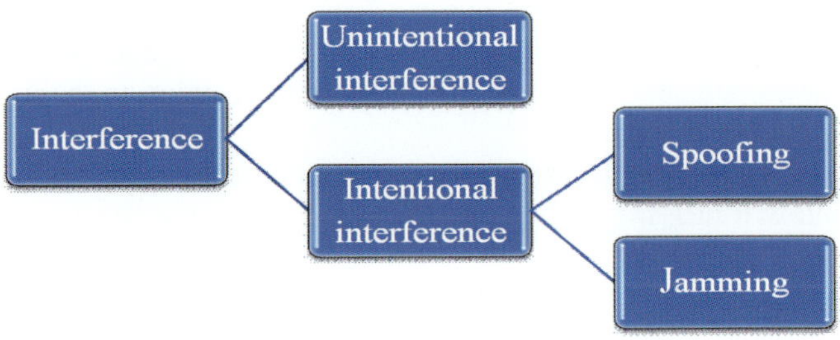

FIGURE 2 Interference types.

Interference Sources

Unintentional Interference Sources

Unintentional interference sources consist of three main components:

1. Other systems in the same frequency band (including other navigation systems with spectrum overlap, transmitters occupying the navigation frequency band illegally).
2. The harmonic, intermodulation, and dispersion components of the signals produced by radio broadcasting and communication emitters.
3. The leakage of electromagnetic radiation in navigation frequency band produced by electronic equipment.

Harmonics may interfere with a variety of civilian GPS band signals as shown in Table 1.

Intentional Interference Sources

The intentional interference usually comes from man-made jammers. At present, civil band jammers can be divided into two types: (1) a suppressing jammer,

TABLE 1 Harmonics May Interfere with GPS L1 Band

ORDER	BAND (MHz)	USAGE
L1	1571.42 - 1579.42	C/A-GPS
2th	785.71 - 788.71	UHF TV
3th	523.807 - 526.473	UHF TV
4th	392.855 - 394.855	Mobile/Station
5th	314.284 - 315.884	Mobile/Station
6th	261.903 - 263.237	Mobile/Station
7th	224.488 - 225.631	Broadcasting
8th	196.427 - 197.428	VHF TV
9th	174.602 - 175.491	VHF TV
10^{h}	157.142 - 157.942	VHF Maritime
11th	142.856 - 143.584	VHF Military
12th	130.952 - 131.618	VHFCOM
13th	120.878 - 121.494	VHFCOM
14th	112.244 - 112.816	VOR/ILS
15th	104.761 - 105.295	FM
16th	98.214- 98.714	FM

Source: Landry and Renard, 1997. Reprinted with permission.

which makes the receiver unable to output the PNT results, and (2) a spoofer, which induces the receiver to give false results.

Suppressing jammers' detailed schemes are freely available from the Internet, and they are low-cost and simple designs. Therefore, the majority of jammers on the market are suppressing jammers. The detailed scheme of a 0.5 W GPS L1 suppressing jammer from the Internet is shown in Figure 3; the vehicle GPS jammer and GPS/ GLObal NAvigation Satellite System (GLONASS) jammer are shown in Figures 4 and 5. Figures 6 and 7 show the low-power and high-power GPS mobile integrated jammers.

Spoofers are more complex and expensive. Common products on the market are the frequency-sweep traction spoofers. The production of spoofers is not easy because of technical difficulties and high cost. Three kinds of spoofers are now being studied (Humphreys et al., 2008). Because the implementation costs are different, the effects of spoofers are not alike. A GPS signal generator is shown in Figure 8 that can be used as the source of interference, but the price is very high.

Because the intentional interference is deliberately caused by human factors, its location, transmission power, and boot time are more variable and difficult to examine.

CIVIL SIGNAL INTERFERENCE TYPICAL CASES

Recently, the news media has reported many civil GPS signal interference cases, and the construction of the BD system has encountered interference. The following will give some relevant cases.

It turns out that the second harmonic of the carrier of a U.S. National Time Service Center Channel 66 television transmitter falls right in the middle of the 1575.42 MHz GPS Band.[1] Although Channel 66 has reduced the coverage, reduced the receiver power to 1,000 watts, and made the harmonic suppression to meet the requirements of the Federal Communications Commission, the GPS receiver is still unable to work within several miles of the transmitter.

In late 2009, engineers noticed that satellite-positioning receivers at Newark airport in New Jersey were suffering brief daily breaks in reception. Something was interfering with the signals from orbiting GPS satellites. It took two months for investigators from the Federal Aviation Administration to track down the problem: a driver who passed by on the nearby New Jersey Turnpike each day had a cheap GPS jammer in his truck.

In late 2009, researchers found that the B2 band of BD system Monitor Station reference receivers had been interfered with and was unable to obtain complete dual-frequency observations. After monitoring, the radio administration bureau found that a company warehouse nearby installed wireless video

[1] Comment on TV Channel 66, "Transmitters Jamming GPS Signals" [EB/OL], at *http://blockyourid.com/~gbpprorg/mil/gps/gps_jammer2.txt*.

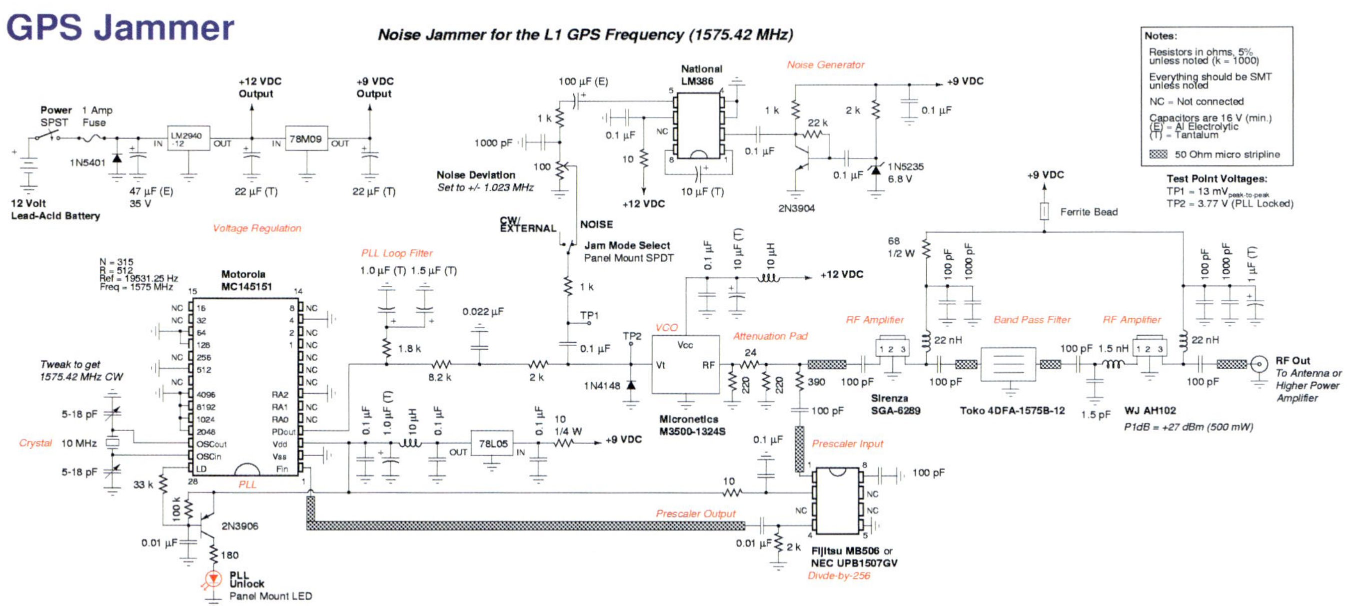

FIGURE 3 Detailed scheme of a 0.5 W GPS L1 suppressing jammer from the Internet, at *http://servv89pn0aj.sn.sourcedns.com/~gbpprorg/mil/gps/gps_jammer.png*.

FIGURE 4 Vehicle GPS jammer.

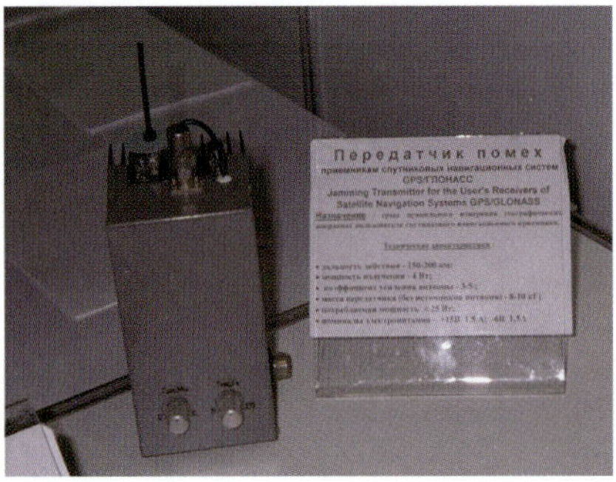

FIGURE 5 GPS/GLONASS jammer.

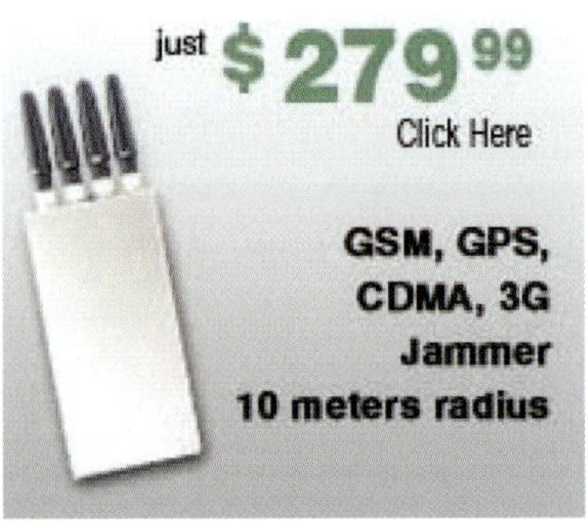

FIGURE 6 Low-power GPS mobile integrated jammer.

FIGURE 7 High-power GPS mobile integrated jammer.

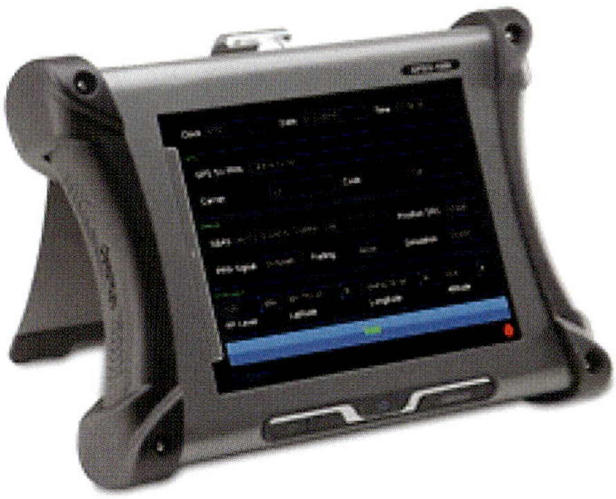

FIGURE 8 Portable GPS signal generator.

surveillance equipment illegally, and the frequency of the emitter occupied the B2 working frequency band.

During 2009–2010, researchers found that the BD system Monitor Station reference receivers had been jammed occasionally in some areas and could not easily determine the interference sources. The receivers recovered after engineers started the built-in adaptive anti-jamming algorithm. Figure 9 displays the jammed power spectrum of B1 band.

As the GNSS has a wide range of applications, the interference would have serious implications. In the fields of traffic, power, communication, and finance, the interference will cause socio-economic disorder. It also poses a threat to armored cars, dangerous goods vehicles, and large equipment. Because of the impact of interference, the receiver cannot locate the position correctly or identify output error. Therefore, the civil aviation landing systems, which rely on GNSS, may lead to fatal catastrophic incidents.

CIVIL SIGNAL INTERFERENCE COUNTERMEASURES

Strengthen Navigation Frequency Spectrum Monitoring Within the Framework of Law

In China, the usage and protection of radio spectrum resources and the investigation, production, distribution, and importation of the radio equipment have clear legal provisions. Relevant laws and regulations are: *Real Right Law of the People's Republic of China*, *Criminal Law of the People's Republic of China*, *Radio Regulations of the People's Republic of China*, and *Radio Station License Regulations.*

The protection of navigation signal spectrum resources relies on both the support of technology and the effective implementation of management measures.

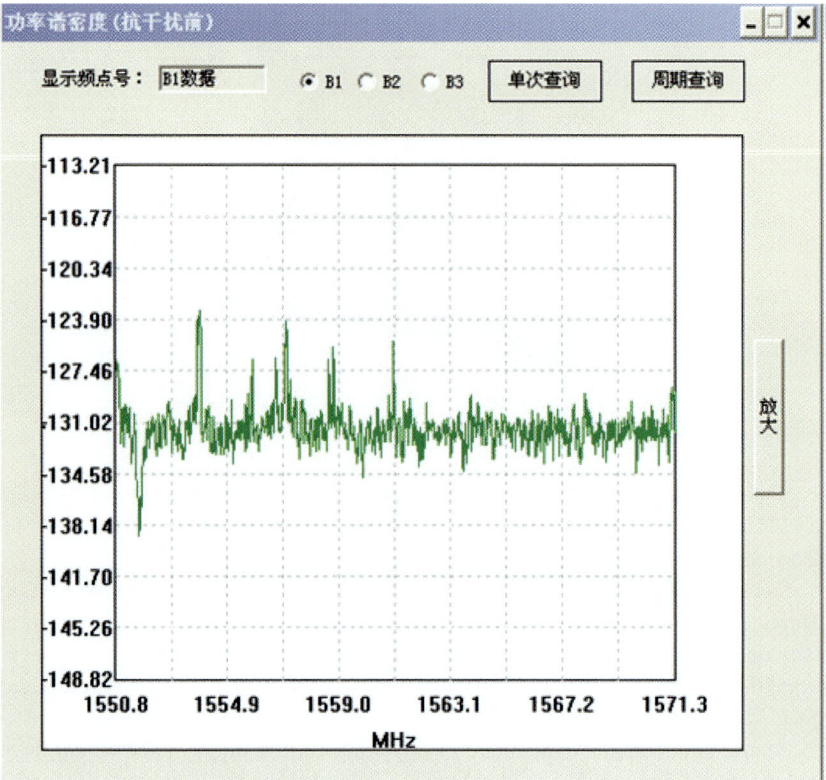

FIGURE 9 Jammed power spectrum of B1 band.

The China Bureau of Radio Regulation of the Chinese Ministry of Industry and Information has overall responsibility for administration and coordination of civil radio management. Its responsibilities are frequency spectrum monitoring, interference detection and investigation, matters of electromagnetic interference coordination, and transmission of radio waves in the air maintenance. Figures 10 and 11 show some radio monitoring equipment.

In the field of radio monitoring, the detection and location of high-power jammers has been a mature technology and has been widely used in engineering. At present, China possesses the ability of radio frequency spectrum detection and interference location investigation. But the low-power interference, which has serious impact on the navigation signals, can only be detected in limited range. The location and detection of a weak jammer takes a lot of manpower, material resources, and time. To enhance the navigation frequency monitoring and the detection of interference sources, the main issues include (1) high-sensitivity interference detection and (2) rapid location of the interference. However, the

FIGURE 10 Fixed radio monitoring stations. Source: Beijing Radio Administration Bureau, at *http://www.bjrab.gov.cn/jcjc/jcs/twzs/20101221/368.shtml*.

FIGURE 11 Radio monitoring vehicles. Source: Beijing Radio Administration Bureau, at *http://www.bjrab.gov.cn/jcjc/jcs/twzs/20101221/364.shtml*.

method of location and removal cannot meet users' requirement of continuous services. Thus receivers with higher interference tolerance are demanded.

Enhance the Receivers' Interference Tolerance

Fully Exploit Anti-jamming Performance of Modernized Navigation Signals

The modernized navigation signals enhance the civil signals' interference tolerance. The anti-jamming potential of modern navigation signals is based on the following aspects.

1. **Integrated application of multiple frequencies to avoid possible navigation unintentional interference.** In the future, the civilian frequencies that can be used include BD-B1, B2, B3, GPS-L1, L2, L5, and GALILEO-E1, E5, E6, etc. Applying multiple navigation frequencies can effectively avoid the impact of unintentional interference, meaning that interference with one frequency signal does not affect the other frequency signals. This is equivalent to increased tolerance. The deficiency of this method is that it is only effective on unintentional interference.

2. **Use of wider bandwidth signal to get greater gain of the spread spectrum.** In the modernized GNSS open signals, there are signals with wider bandwidth, such as BD-B2 and GPS-L5. The higher chip rate increases the processing gain and anti-jam performance. So it can improve the ability of anti-jamming. Such a measure is both effective on intentional and unintentional interference.

3. **Pilot signals improving tracking capability.** Pilot signal is modulated by long-period PRN code without navigation message. Receivers can extend the coherent integration time as long as possible to improve signal-to-noise ratio and eliminate the impact of square loss, thus the sensitivity can be significantly improved. So it can enhance the robustness of the equipment to the interference condition. Such a measure is both effective in intentional and unintentional interference.

4. **New message error correction coding scheme.** Modernized GNSS signals adopt new-style forward error correction coding schemes. High-gain, soft-decision decoding can correct the navigation message error and improve the continuity of navigation in various kinds of interference. Such a measure is both available in intentional and unintentional interference.

Improving Anti-jamming Capability through Auxiliary Information

Currently, satellite navigation has been closely integrated with inertial navigation and mobile communication. In the future, integration with other systems will become a trend. Full use of auxiliary information of other systems or equipment will greatly improve the anti-jamming capability of satellite navigation receivers. At present, the method can be used in the following aspects:

1. For static timing users, the receivers' interference tolerance can be enhanced by setting precise coordinates and using high-stability clock.
2. For vehicles, aircraft, and other dynamic users, GNSS/inertial navigation system integration can improve anti-jamming capability.
3. With messages and other necessary information broadcast by 2G or 3G mobile communication networks, pilot positioning can be achieved directly and anti-jamming capability can be improved.

Active Anti-jamming Measures

Adding active anti-jamming measures to receivers can upgrade the ability of the anti-jam. At present, the methods that can be used in active anti-jamming include:

1. **Time and frequency domain anti-jamming.** Time and frequency domain processing can suppress various narrowband interference effectively. Anti-jamming capability can be improved by 40–70 dB based on the architecture of hardware and processing algorithms. The main advantages of the approach are that the instruments are low cost and easily integrated; however, its disadvantage is it is ineffective to wideband interference. Figure 12 presents the power spectrum comparison before and after anti-jam.

2. **Spatial anti-jamming processing.** Spatial processing is mainly using the antenna array disposal techniques from the direction of signal to distinguish the interference. Adaptive space-frequency and space-time filtering technology can effectively suppress various types of interference. According to the hardware scale and processing algorithms, we can get 30–60 dB or higher anti-jam improvement. Spatial processing can suppress multiple wideband and narrowband interference at the same time. But it has the disadvantages of complex hardware, large scale, and high cost, and the number of anti-wideband interference is directly related to the scale of hardware and software. Figure 13 presents the directional map of antenna array in spatial processing.

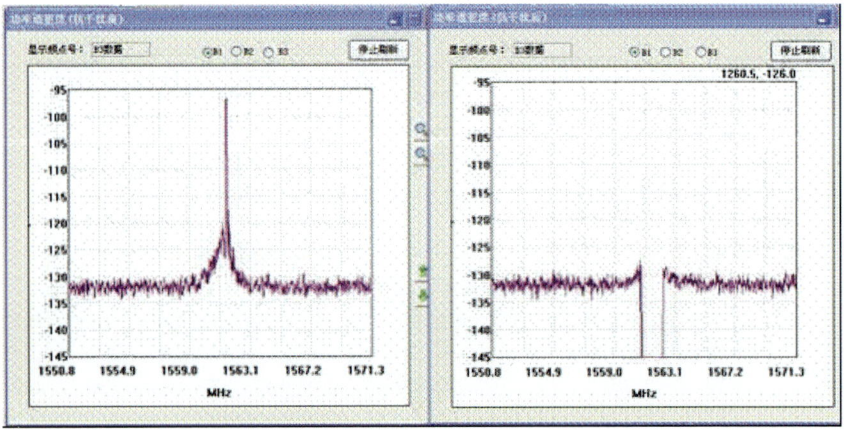

FIGURE 12 Power spectrum comparison before and after anti-jam.

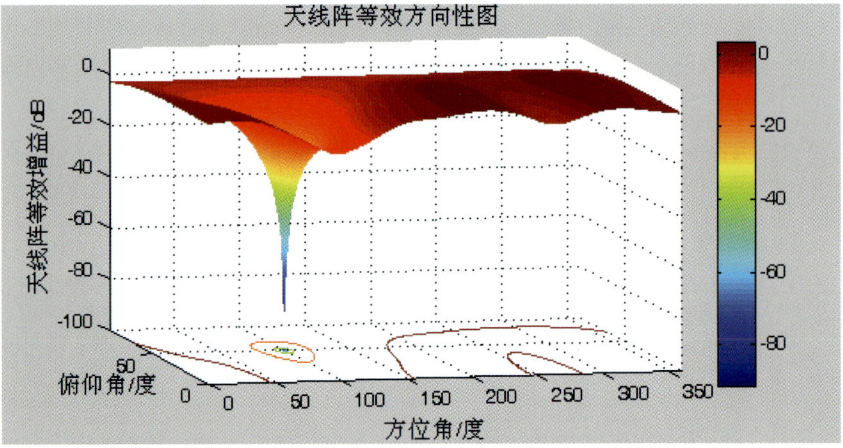

FIGURE 13 Directional map of antenna array in spatial processing.

In order to increase the active anti-jamming measures, we have to increase the complexity, size, power consumption, and implementation costs of the receivers. However, engineering has to considerer cost-effectiveness. For the individual user equipment, because of its constraints on size and power, we propose to use time-domain to suppress narrowband interference. For the critical civilian infrastructure equipment, we can appropriately increase the hardware scale and use the spatial-domain or polarization to enhance the anti-jamming capability and protect the safety of infrastructure.

Developing the Auto-integrity Technology in the Background of Interference

Using the RAIM (receiver autonomous integrity monitoring) technology, receivers can monitor abnormal navigation signals and mitigate errors of PNT calculation caused by false signals.

SUMMING UP

- Governments have a responsibility to protect the navigation frequency band against illegal interference.
- Technical measures can greatly improve the receiver's anti-jamming capability. Good PNT services can be obtained by adopting proper anti-jamming techniques according to the role of GNSS in specific applications.
- Active anti-jamming technology will affect the accuracy of measure precision. Therefore, in the mapping and other precision application fields, active anti-jam technology is limited.
- Interference-free environments are the goal pursued by all satellite navigation users.

REFERENCES

Landry, R. Jr., and A. Renard. 1997. Analysis of Potential Interference Sources and Assessment of Present Solutions for GPS/GNSS Receivers[C]. 4th Saint-Petersburg on INS, May 26–28.

Humphreys, T.E., B.M. Ledvina, M.L. Psiaki, B.W. O'Hanlon, and P.M. Kintner Jr. 2008. Assessing the Spoofing Threat: Development of a Portable GPS Civilian Spoofer[C]. ION GNSS Conference Savanna, Georgia, September 16–19.

Present and Future Applications of COMPASS Navigation Satellite System

TAN SHUSEN

ABSTRACT

After an overview of the COMPASS Navigation Satellite System, the author introduces its Radio Determination Satellite Service (RDSS) and Regional Navigation Satellite Service (RNSS) and their performances, which can serve group users based on location information sharing. The typical application modes of COMPASS group users are presented. Furthermore, possible applications in the near future are also explored.

OVERVIEW OF COMPASS'S THREE-STEP DEVELOPMENT STRATEGY

Research on satellite navigation in China was first initiated in the 1980s. Mainly learning from the GPS and GLObal NAvigation Satellite System (GLONASS), China had been groping for a development strategy for its own satellite navigation system. It is Academician Chen Yunfang who first proposed a positioning theory that makes use of two geostationary Earth orbit (GEO) satellites and user elevation. In 1994, the project based on this idea was formally approved by the state. A demonstration system of COMPASS was established after the successful launch of three GEOs between 2000 and 2003.

The COMPASS demonstration system consists of three GEOs, all of which are equipped with RDSS payloads (Figure 1). One of them is also equipped with RNSS experimental payloads. RDSS is the major service, with functions of positioning, user location report, short message communication, and timing.

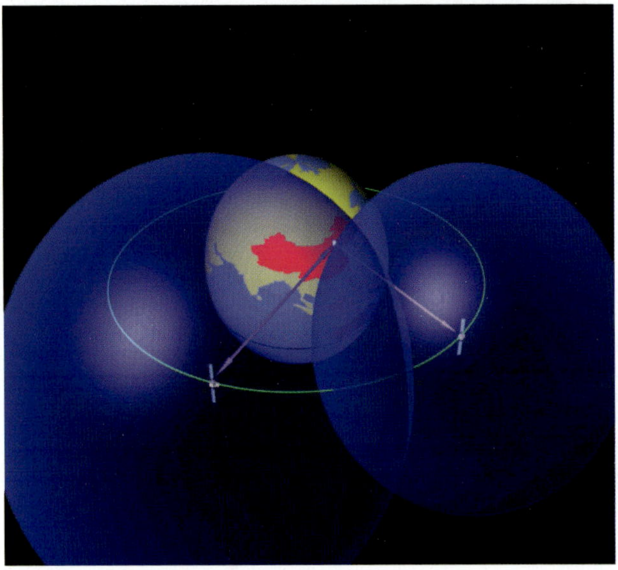

FIGURE 1 COMPASS Navigation Satellite System.

Phase II of COMPASS aims at meeting users' demands in the Asia-Pacific region (Figure 2). A constellation made up of 12 satellites will be put in place before 2012. By then, the COMPASS system with RDSS and RNSS will be able to provide users with continuous positioning, velocity measurement, and location report services. COMPASS will achieve compatibility with GPS and provide equivalent accuracy as GPS.

Phase III of COMPASS is to establish a global constellation with 30 to 35 satellites by 2020 (Figure 3). It adopts RDSS and RNSS as well. As a member of the GNSS community, COMPASS will achieve compatibility and interoperability with GPS and Galileo within multi-frequency bands.

COMPASS SYSTEM'S PERFORMANCE

RDSS and RNSS

RDSS can provide information of both the user's location and time parameters (X, Y, Z, T). It can provide location reports among the users, short message and timing services at the same time.

RNSS can provide user's location, velocity, and time parameters (X, Y, Z, Vx, Vy, Vz, T).

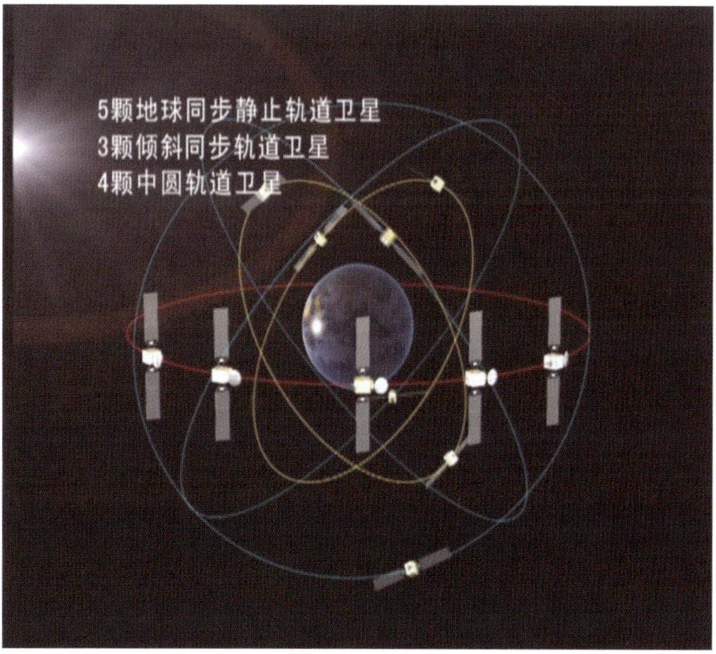

FIGURE 2 Phase II of COMPASS Navigation Satellite System.

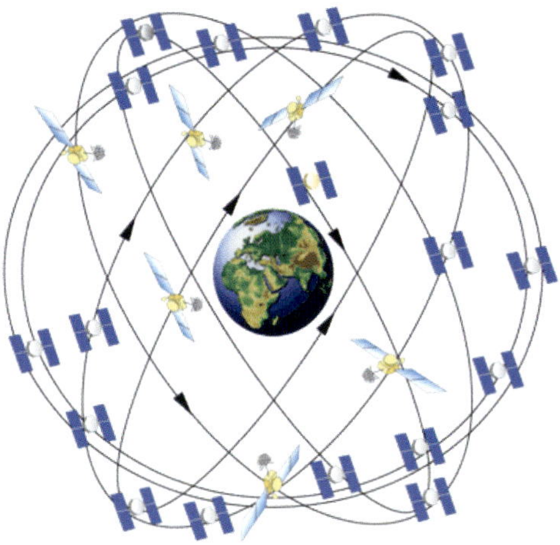

FIGURE 3 Phase III of COMPASS Navigation Satellite System.

Performance

Accuracy levels of the positioning and timing services provided by the COMPASS system can reach 10 m and 20~50 ns respectively, while the response time for its positioning and location reporting is 1 s. Two more features of the system are location information sharing among group users and feasibility of achieving possible interoperability with GPS as well as Galileo.

Major Group Users

The COMPASS system is mainly applied to group users based on location information services in the following sectors:

- Traffic Management, Transport of Hazardous Goods;
- Marine Fishery, Geology, Water Reserve;
- Fleet Management, Forest Fire Prevention, Relief in Earthquake and Flood; and
- Aerial and Offshore Rescue.

APPLICATION MODE

No matter what kind of application of COMPASS, there are three elements: (1) various sensors, (2) the COMPASS information system, (3) and the group user's management center, which can be illustrated as in Figure 4.

A brief explanation of the three elements follows.

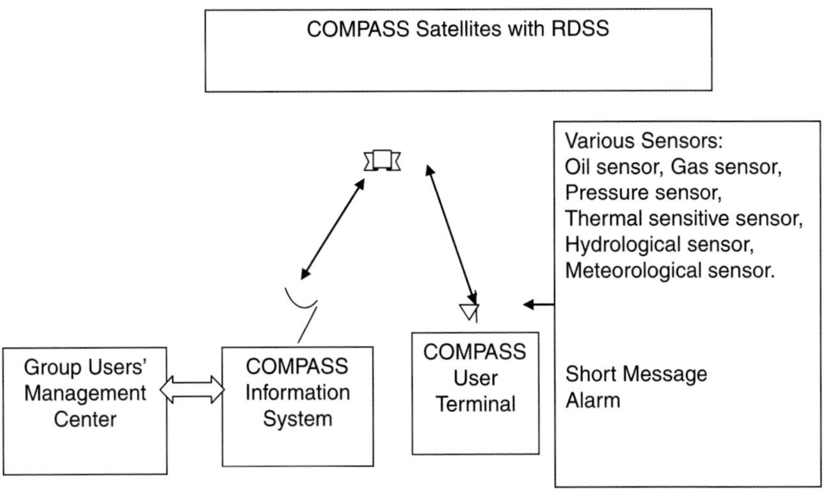

FIGURE 4 Typical application.

First, various sensors will gather information that will be used for further processing, such as the onboard sensors of networked vehicles, as well as onboard oil and gas sensors of networked ships, etc.

Second, the COMPASS information system, usually incorporating the user terminal, RDSS satellites, and COMPASS application center, can achieve marking of moving articles' position and time, data transmission, and distribution.

Finally, the group user management center is in charge of cloud computing, intelligent processing, and real-time dispatching and management in various industries to satisfy the needs of all users within the group.

The above three elements constitute an integrated application mode of COMPASS Navigation Satellite System.

TYPICAL APPLICATION PROJECTS

COMPASS Dispatching System in Marine Fishery

China has realized offshore real-time monitoring of about 20,000 fishing vessels by taking advantage of COMPASS in rapid positioning, location report, and short message communication (Figure 5). Moreover, many other functions, such as alarms for dangerous zone and bad weather, monitoring and commanding, ship tracking, operation track recording, and oil fuel data collection, etc., are possible. An even larger-scale system for offshore group users is under construction.

COMPASS Hydrological Data Collection System at Three Gorges of the Yangtze River

Regular water reports about the Three Gorges Conservatory of the Yangtze River are urgently needed by local governments. The Hydrological Data Collection System (Figure 6) can get reports on the tributaries along the upstream of the Yangtze River covering an area of about 370,000 square kilometers; can provide automatic monitoring and reporting for cascade hydroelectric stations on the downstream of Jinsha River, with a coverage of 45.443 square kilometers; and can report on the hydroelectric stations on the middle stream of the Jinsha River. The system has successfully solved the problems of delays in flood forecasting for the upstream of the Yangtze River and scarce monitoring stations and inconvenient communication.

Moreover, a weather station has been set up at Mount Everest that helped in meteorological observations for the Beijing Olympic Games in 2008, as illustrated in Figure 7.

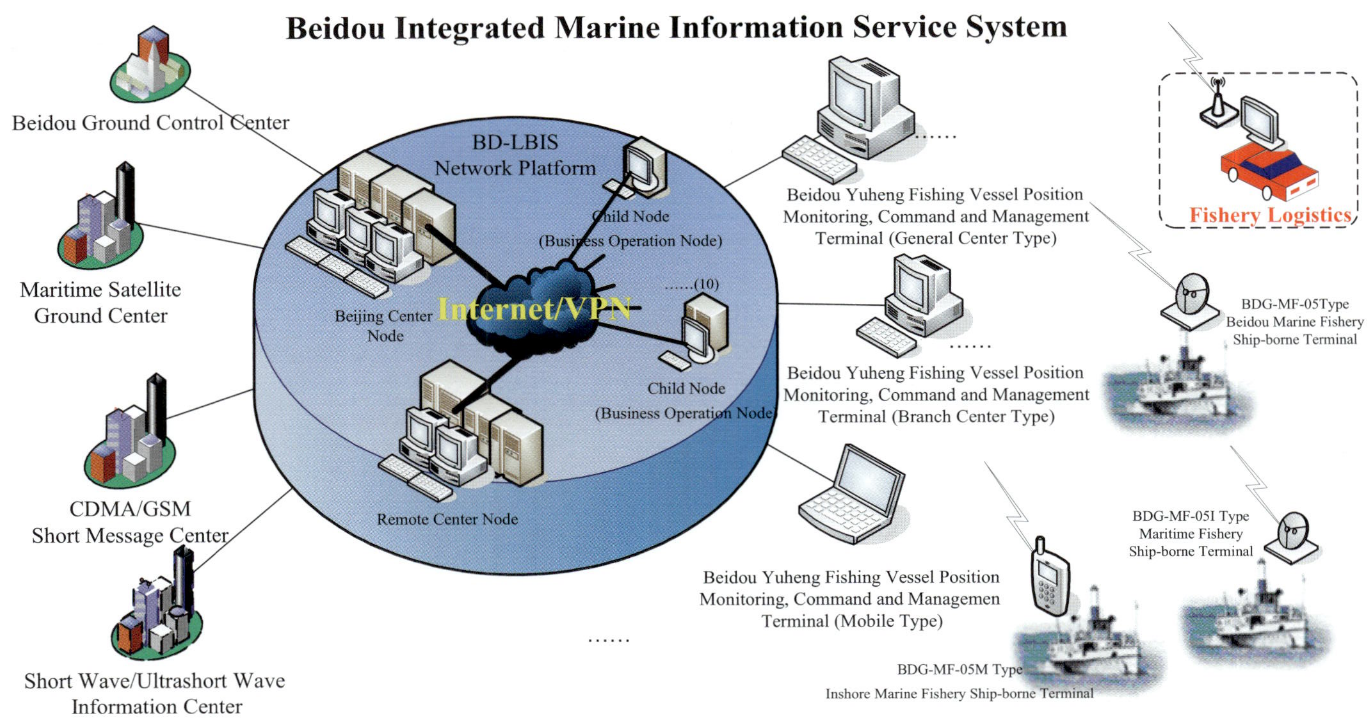

FIGURE 5 COMPASS dispatching system in marine fishery.

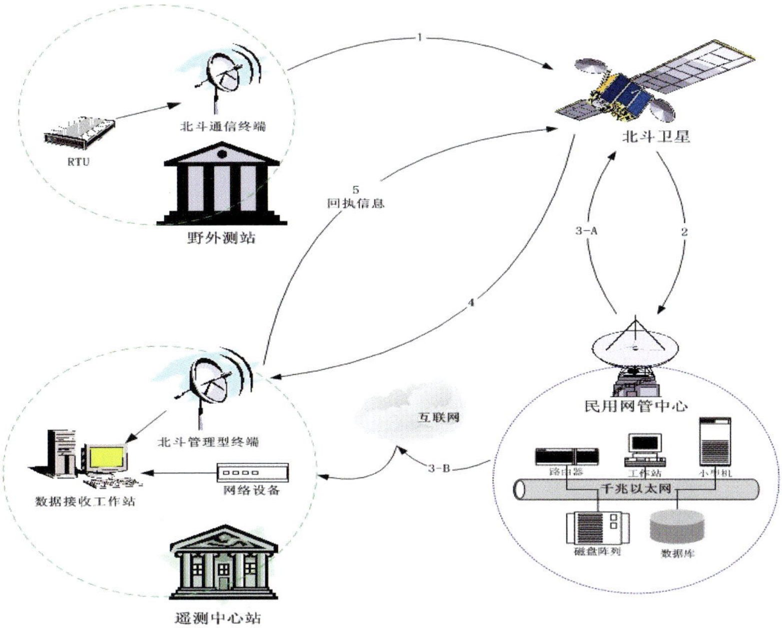

FIGURE 6 COMPASS hydrological data collection system.

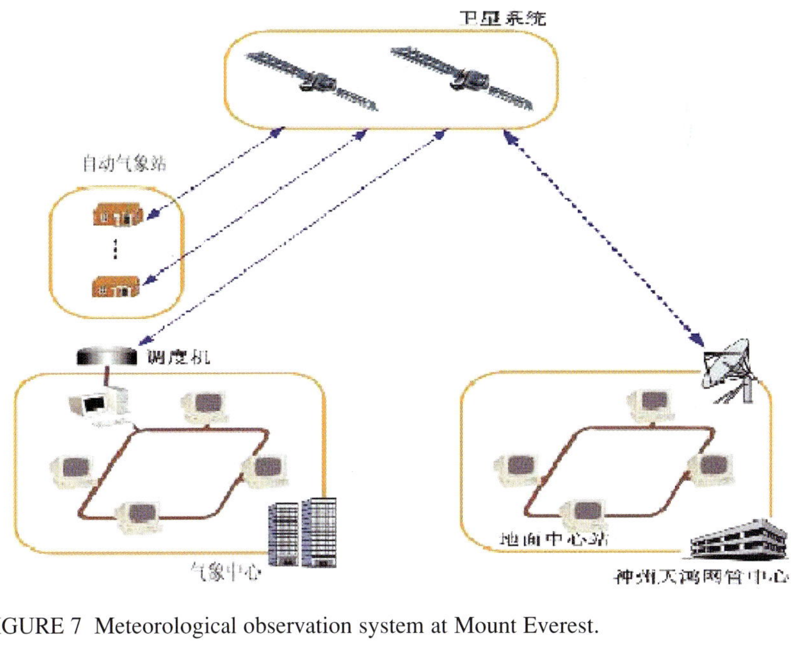

FIGURE 7 Meteorological observation system at Mount Everest.

Detection and Report of Forest Fire by COMPASS

A forest fire warning system based on COMPASS is supported by the National Forestry Bureau. This system can send real-time positions of all fire fighters, vehicles, and airplanes, make the related digital map, and keep everyone in contact by short messages. In this system, fire fighting airplanes can outline the areas on fire and inform the rescuers regarding tree species, terrain, and the onsite situation of the on-fire area, which is illustrated in Figure 8.

Inspection and Monitoring System for High-voltage Power Line in Remote Areas

The inspection and monitoring system for the 500-kilovolt power transmission line, which extends about 3,000 kilometers in total and has 4,000 inspection

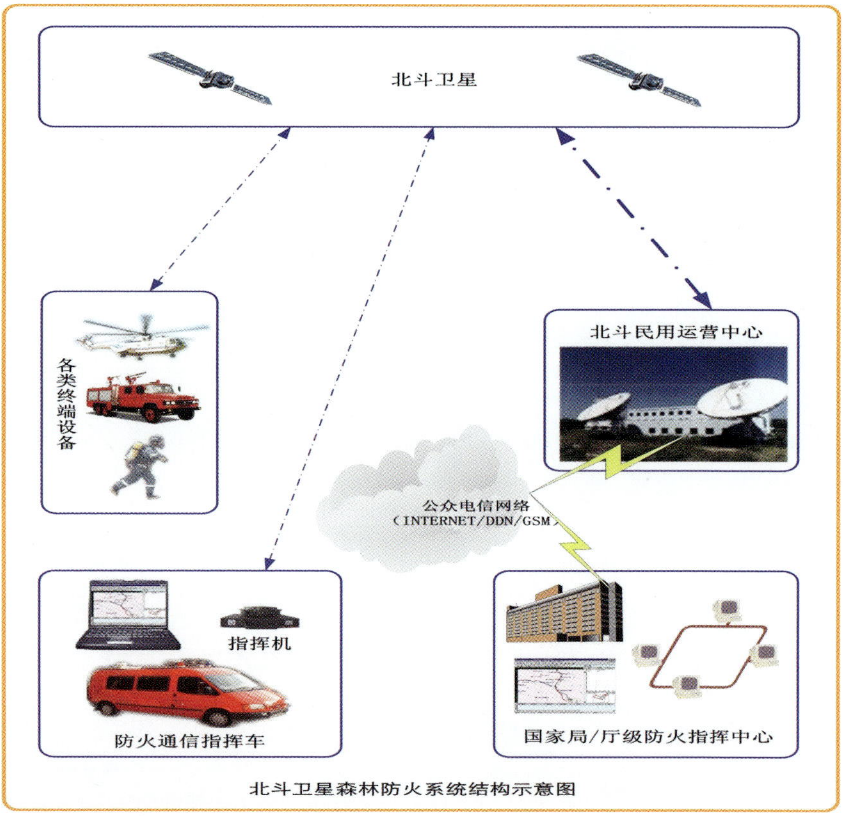

FIGURE 8 COMPASS forestry fire prevention system.

tower-poles in the Tsitsihar Super-grid Bureau of the State Grid Cooperation of China, can provide real-time inspection, dispatching, security, and recording of staff, vehicle, and helicopter. The inspection and monitoring system used by helicopter for high-voltage power lines is shown in Figure 9.

Yangtze River Environmental Protection and Energy Conservation in Shipping

COMPASS performs as a data center as well as distribution center in environmental protection and energy conservation of the Yangtze River shipping.

With the help of COMPASS, this vessel monitoring system provides real-time monitoring of the on-board and offshore dispatching systems. In this way, not only can voyage security be guaranteed but also exhaust emissions and energy

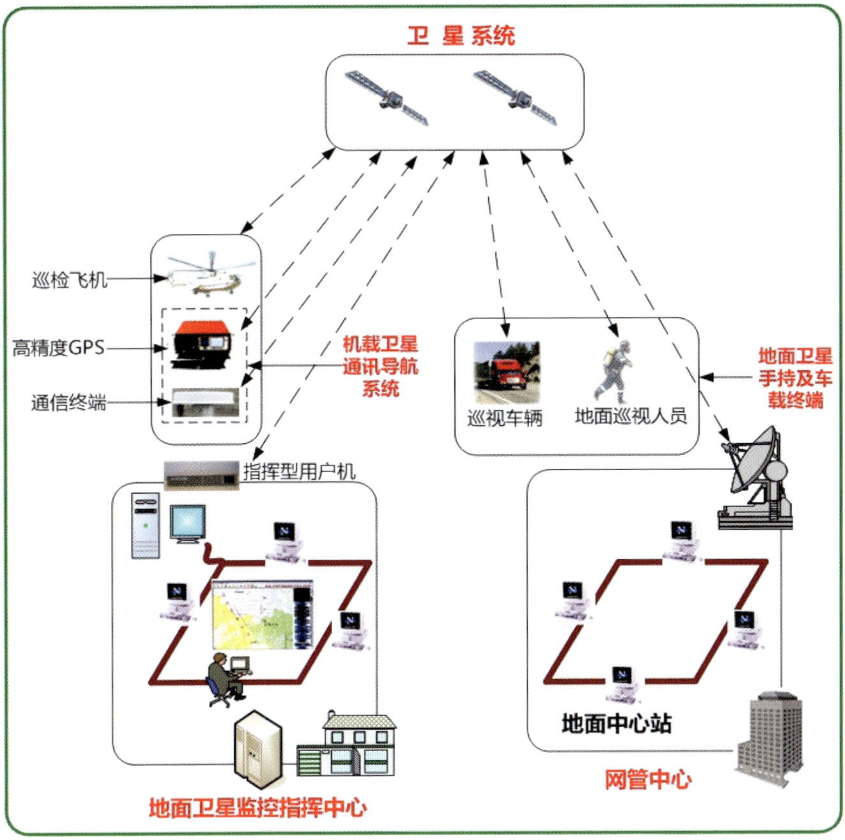

FIGURE 9 Inspection and monitoring system for high-voltage power line.

consumption can be reduced. The latter, however, is of more important significance to us.

COMPASS Vehicle Navigation System

The main objective of an automobile network is to share information of all vehicles in operation and to help lessen traffic jams. The key lies in the data transmission of cars on the road to the information center. COMPASS RDSS performs as a radio communication and positioning system, which is like the first runner in the automobile network relay race. The vehicle navigation system is illustrated in Figure 10.

The Two-Way Timing Service of COMPASS Can Provide a Standard of High Accuracy Time and Frequency for Upgrading the Old Power Grid

On February 18, 2011, an industrial park for COMPASS satellite navigation civil applications was set up at the Jiaozhou City of Shandong Province. Initial results of COMPASS timing services for the power industry have been satisfactory. Also in 2010, intelligent upgrading of 23,500-kilovolt stations of the East China Power Grid was completed successively. By taking advantage of high-accuracy timing services provided by COMPASS, both the accident rate caused by timing accuracy and the malfunction rate of sampled facilities are zero, exceeding the internationally accepted standards. How the time synchronization system of the intelligent power grid is incorporated is shown in Figure 11.

PROSPECTS OF INTEROPERABILITY BETWEEN COMPASS AND GPS

Coordination Between COMPASS and GPS Under the ITU Framework Has Been Accomplished

In September 2010, frequency coordination between COMPASS and GPS on the L band was finished, with the frequency compatibility of both authorized and open signals. Consensus has been achieved on the maximum and minimum power of B1C/L1C and B2a/L5 as well as the sequential correlation of spreading codes.

Interoperability

Interoperability in B1C/L1C and B2a/L5 will be discussed for the users' benefits to the largest extent.

Enormous efforts will be made to reduce complexity of receivers and provide the most convenient access.

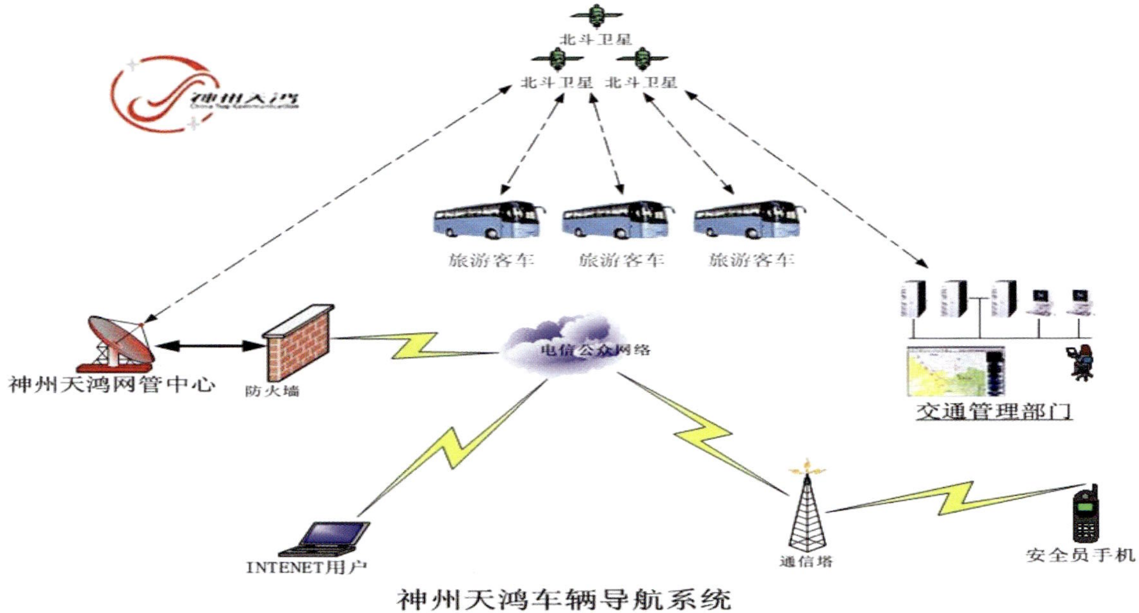

FIGURE 10 Vehicle navigation system.

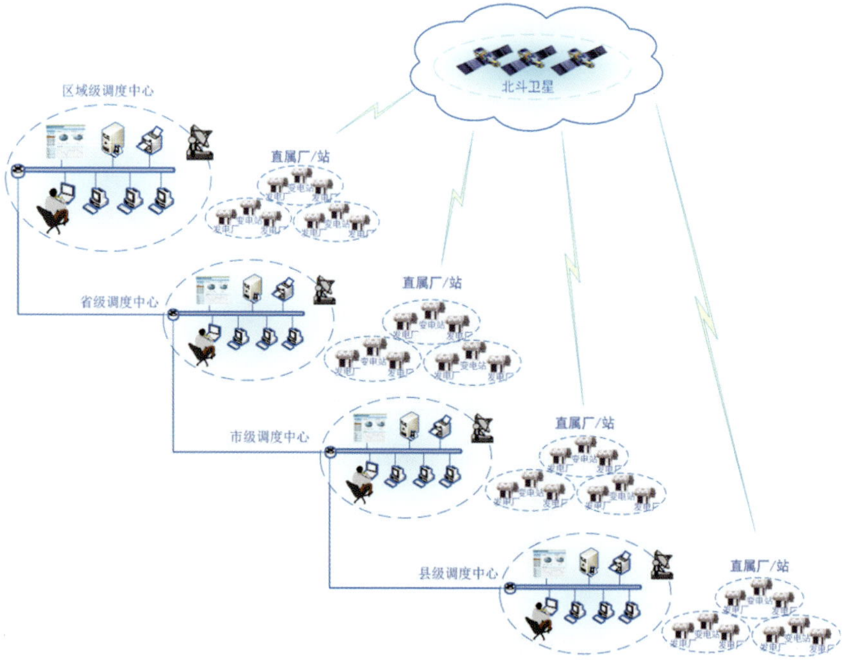

FIGURE 11 COMPASS two-way timing service system.

CONCLUSION

1. COMPASS has been largely used as the three-step development strategy planned.
2. The position data sharing service by COMPASS has been widely accepted.
3. There is a good and broad foundation for cooperation in compatibility and interoperability for GPS and COMPASS.
4. Some Perspectives on Future Cooperation

There are many possible areas of cooperation between COMPASS and GPS, such as aerial navigation application and standards discussion, geodetic surveying and the coordinate system, receiver technology and consumer products, as well as mitigating the interference with civil signals, etc.

Application of GNSS to Environmental Studies

PENINA AXELRAD
Colorado Center for Astrodynamics Research
University of Colorado at Boulder

ABSTRACT

GNSS signals are influenced by the transmission media and interaction with surfaces near the receiving antenna. Observations of the modified signals from the ground and from airborne and spaceborne platforms allow for scientific study of the ionosphere, atmosphere, and Earth surface. Methods using standard ground-based receivers provide estimates of atmospheric water vapor and soil moisture. Specialized receivers measuring occulted signals enable high-resolution estimates of atmospheric density and temperature. Receivers measuring reflected signals are used to infer surface roughness and reflectivity, which can be related to surface conditions like ocean winds, soil moisture, and ice type. Modern receivers that can use multiple GNSS constellations will provide a rich global data set for environmental study.

INTRODUCTION

Intended primarily for position, navigation, and timing (PNT), Global Navigation Satellite Systems (GNSS) bathe Earth with a multitude of highly stable ranging signals that are readily available for use in probing the atmosphere and surface of Earth. Furthermore, GNSS receivers deployed in diverse environments (land, marine, air, space) for conventional PNT purposes can also be utilized to make environmental observations with these same signals. All GNSS signals received on the ground or in low Earth orbit traverse the *ionosphere*. At L-band, frequency dependent dispersion of modulation and carrier phase allows total electron content to be observed by measuring range and/or phase at two or more frequencies, or through the dispersion of code and carrier. *Atmospheric* attenuation and delay observed by ground stations can be modeled/observed by estimating parameters associated with empirical models. Measuring the phase delay and amplitude variations of occulting signals from aircraft or satellites enables high-resolution retrieval of atmospheric density, which also exposes temperature and water vapor content. Signals *reflected* by Earth contain information on surface properties including roughness and reflectivity. Measuring reflected signals from ground, aircraft, or spacecraft enables retrieval of surface conditions including soil moisture, sea ice type, and ocean surface winds.

ATMOSPHERIC SENSING

Ground-based receivers at known locations measuring pseudorange and phase to all GNSS satellites in view provide useful observations for estimating precipitable water vapor (PWV) in the atmosphere. The integrated atmospheric effect along the satellite to receiver ray path can be isolated by removing all other sources of error. The technique was established in the early 1990s (Bevis et al., 1992; Rocken et al., 1993) and is done fairly routinely today (Rocken et al., 2005), although the estimates are not sufficiently widespread to be operationally incorporated in the weather prediction models. Ground-based GNSS PWV sensing relies on a network of geodetic-quality dual-frequency receivers at known locations, precise orbit models, and accurate models relating temperature and moisture content in the atmosphere. High-precision software like Bernese (Dach et al., 2007) must be used to ensure that all other error sources have been eliminated.

A more powerful worldwide approach to exploit GNSS for atmospheric sensing relies on radio occultations measured by orbiting satellites. Researchers have studied the atmospheres of Mars, Venus, and Jupiter since the 1970s through radio occultation. The use of GPS for radio occultation (GPS-RO) measurements from LEO was first initiated in 1995 by the GPS/MET experiment (Schreiner et al., 1998; Yunck et al., 2000). This satellite, placed in a 735 km orbit at 70 degree inclination, flew a modified TurboRogue receiver designed to track signals from above for precise orbit determination and from a separate antenna, signals passing close to the limb of Earth. Now, in 2011, there are about a dozen satellites flying operational or experimental occultation payloads including the six-satellite COSMIC constellation (Anthes et al., 2008), CHAMP, SAC-C, GRAS/Metop, C/NOFS, GRACE-A, and TerrSAR-X. These platforms use forward and backward facing antennas to measure GNSS signals as they rise or set and are occulted by Earth's atmosphere as shown in Figure 1. A vertical profile of the bending angle through the ionosphere, stratosphere, and troposphere is determined from the excess phase measurements. Refractivity is derived from the bending angles and then further analyzed to determine electron density profiles, temperature, pressure, and water vapor. At its peak performance, COSMIC provided approximately

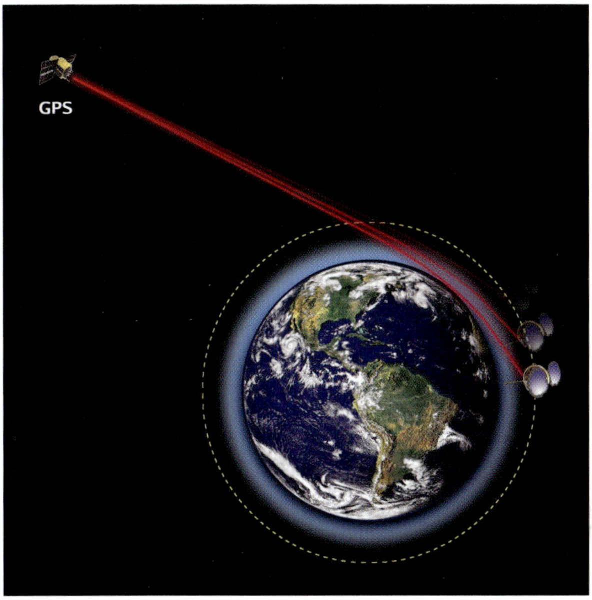

FIGURE 1 Illustration of GPS radio occultation. Sources: The Earth image is provided by the SeaWiFS Project, NASA Goddard Space Flight Center, and ORBIMAGE. The Leo satellite is "Courtesy of Orbital Corporation." The image composition is © University Corporation for Atmospheric Research. Courtesy University Corporation for Atmospheric Research.

2,500 soundings per day. These observations are already being used as key inputs to numerical weather prediction (Cucurull et al., 2007), atmospheric studies, and climate monitoring, and their impact and significance are expected to grow in the future (Anthes et al., 2008). Compelling results have also been presented (Huang et al., 2010) illustrating the positive benefit of GPS occultation observations on severe weather prediction. New applications being developed include measurement of the planetary boundary layer, temperature inversions, and turbulence. It is notable that a National Research Council decadal survey in 2007 recommended that GPS-RO be made operational by NOAA and that additional GPS-RO instruments be put on NASA science platforms whenever appropriate (NRC, 2007).

Increasing the number of GNSS transmitters and the constellation properties with Compass and Galileo will directly increase coverage of Earth's atmosphere and the timeliness of the observations. This will have a tremendous impact on understanding of Earth in the long term (climate change) and short term (numerical weather prediction, severe storm monitoring). Adding new LEO observing satellites able to track multiple GNSS systems will contribute thousands of observations daily.

SURFACE SENSING

When GNSS is used for conventional PNT purposes, reflections from surfaces near the receiving antenna, termed multipath, introduce errors in the solution by distorting the composite signal tracked by the receiver. By deliberately monitoring and modeling the effects of the reflections, surface and near-surface properties can be inferred. Two basic approaches are being used. The first, pioneered by Kristine Larson (Larson et al., 2008a, 2008b, 2010; Small et al., 2010) uses standard geodetic ground-based receiving equipment and models the effect of the surface reflections on the composite signal tracked by the receiver. The second approach, initially investigated by Martin-Neira (1993) and separately by Katzberg and Garrison (1996) uses an airborne or spaceborne receiver with both a standard upward-facing and specialized downward-facing antenna to separately track direct and reflected signals.

Figure 2 shows the oscillations in the signal-to-noise ratio that are typical of GPS tracking in the presence of ground reflections at GPS tracking sites like the Plate Boundary Observatory (PBO) site in Marshall, Colorado (Figure 3). By relating the frequency and phase shift of these oscillations to the reflectivity and location of the reflection point on the ground, researchers have been able to

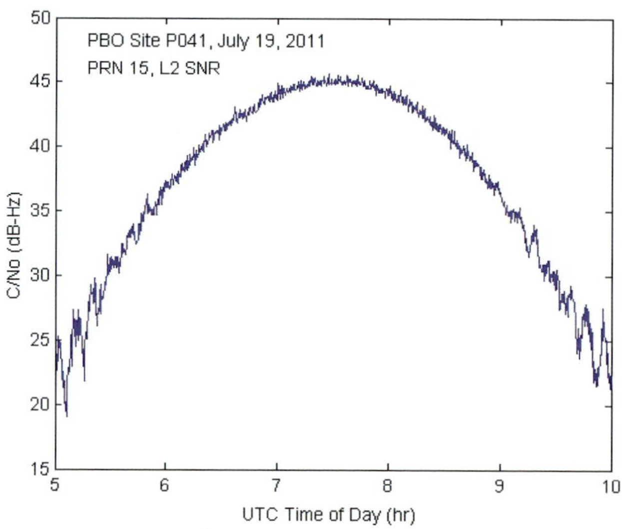

FIGURE 2 Time history of C/No from Marshall, Colorado, PBO site. Oscillations near the start and end of the pass are due to multipath from the ground. Note: This plot is based on data provided by the Plate Boundary Observatory operated by UNAVCO for EarthScope (*http://www.earthscope.org*) and supported by the National Science Foundation (No. EAR-0350028 and EAR-0732947).

observe changes in soil moisture content (Figure 4) (Larson et al., 2008a, 2008b, 2010), crop growth (Small et al., 2010), and snow depth (Larson et al., 2009). Because this approach relies on standard geodetic receivers, the proliferation of such receivers for other geodetic purposes including earthquake monitoring, will make it possible to coincidentally observe Earth surface conditions with no additional expense or installations.

FIGURE 3 GPS Antenna at Marshall, CO PBO site used for soil moisture and snow experiments. Source: Larson et al., 2008a. Courtesy UNAVCO.

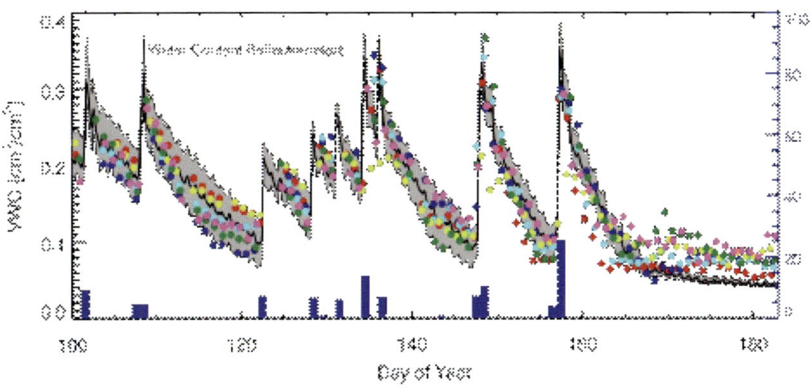

FIGURE 4 Daily precipitation (blue), water content reflectometer range (gray), and GPS soil moisture measurements (colors) for PBO GPS site at Marshall, Colorado. Source: Larson et al., 2008b, Figure 3. Courtesy of Kristine Larson,

Figure 5 shows an airborne configuration for surface sensing using GNSS reflections, and Figure 6 (Esterhuizen, 2006) illustrates the receiver elements for tracking direct and reflected signals for this scenario. For this delay-mapping approach, a zenith antenna receives direct GPS signals, and a nadir antenna (with opposite polarization) receives the same GPS signals after reflection from surface below. Figure 7 compares the signal correlation function for a direct and reflected signal. The delay, attenuation, and spreading of the reflected signals provide the observability of surface properties. Various researchers have demonstrated the ability to detect ocean surface winds (Garrison et al., 2002), soil moisture (Katzberg et al., 2005; Masters et al., 2004), ice and frozen surface roughness (Rivas et al., 2010), and land cover from these observations. While initial results look promising, this technique has not yet been as firmly established as occultation measurements of the atmosphere.

Measurements from space of GPS signals reflected from Earth's surface are also of interest. The concept was first demonstrated from space using data from the Shuttle SIR-C radar (Lowe et al., 2002). Gleason (2006) showed observations of ocean, ice, and land from GPS reflection data collected onboard the UK-DMC.

CONCLUSIONS

Innovative uses for GNSS signals for probing Earth's atmosphere, ionosphere, and surface will certainly continue to develop as more signals and more advanced signals become available. This will enable an ever-increasing improvement in our ability to measure and predict changes in Earth's environment.

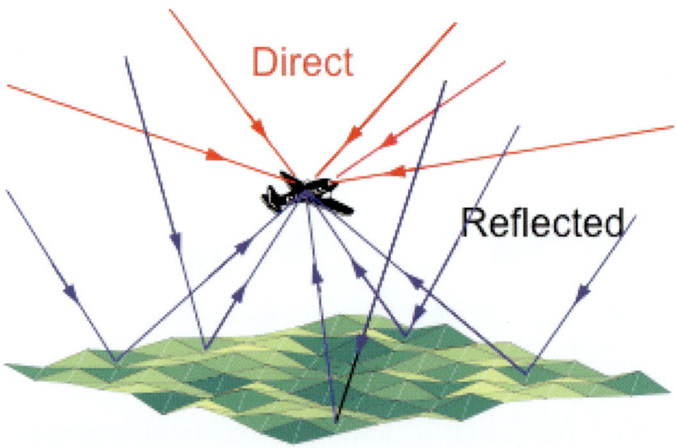

FIGURE 5 Airborne surface sensing based on reflected GNSS signals.

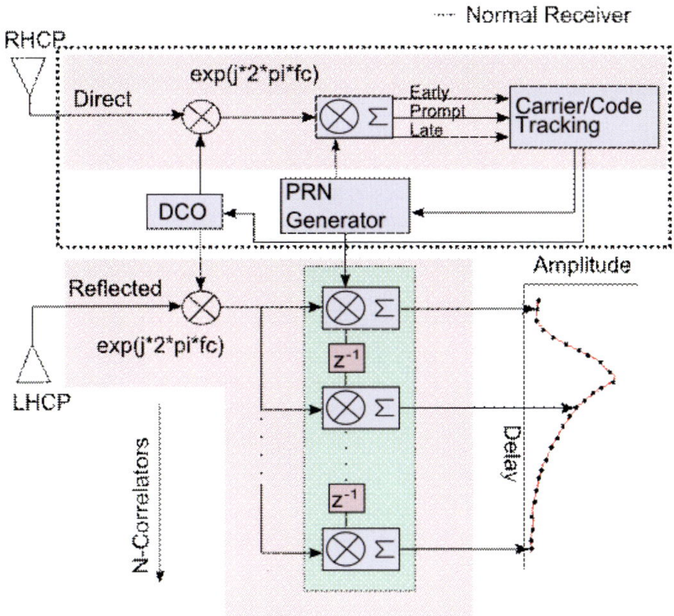

FIGURE 6 Delay mapping receiver used for characterization of surface height, roughness, and reflectivity. Source: Esterhuizen, 2006, Figure 2.5. Courtesy of Stephan Esterhuizen.

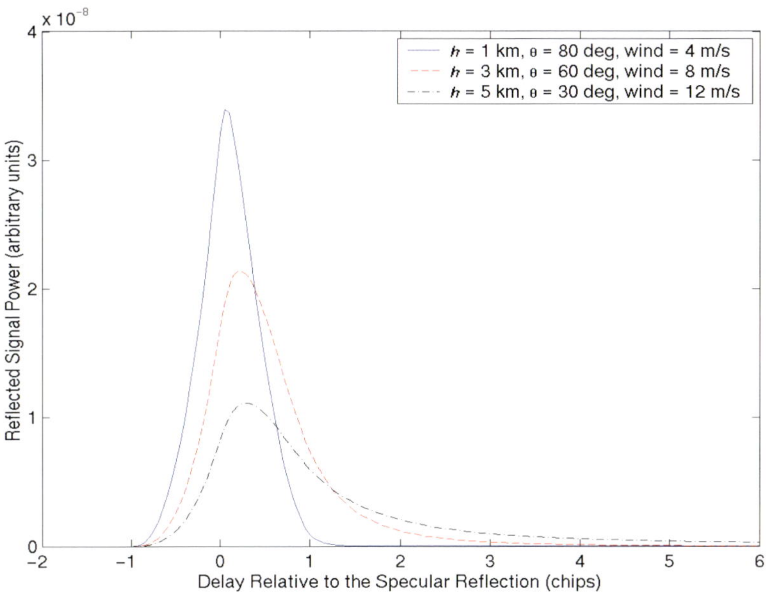

FIGURE 7 Model of effect of surface conditions on correlation function for reflected signals.

REFERENCES

Anthes, R.A., P.A. Bernhardt, Y. Chen, L. Cucurull, K.F. Dymond, D. Ector, S.B. Healy, S.-P. Ho, D.C. Hunt, Y.-H. Kuo, H. Liu, K. Manning, C. McCormick, T.K. Meehan, W.J. Randel, C. Rocken, W.S. Schreiner, S.V. Sokolovskiy, S. Syndergaard, D.C. Thompson, K.E. Trenberth, T.-K. Wee, N.L. Yen, and Z. Zeng. 2008. The COSMIC/FORMOSAT-3 mission: early results. Bulletin of the American Meteorological Society 89(3): 313–333.

Bevis, M., S. Businger, T.A. Herring, C. Rocken, R.A. Anthes, and R.H. Ware. 1992. GPS meteorology: remote sensing of atmospheric water vapor using the Global Positioning System. Journal of Geophysical Research 97(D14): 15787–15801.

Cucurull, L., J.C. Derber, R. Treadon, and R.J. Purser. 2007. Assimilation of Global Positioning System radio occultation observations into NCEP's Global Data Assimilation System. Monthly Weather Review 135(9): 3174–3193.

Dach, R., U. Hugentobler, P. Fridez, and M. Meindl. 2007. Bernese GPS Software Version 5.0 Documentation. Astronomical Institute, University of Bern, Switzerland.

Esterhuizen, S. 2006. The Design, Construction, and Testing of a Modular GPS Bistatic Radar Software Receiver for Small Platforms. M.S. Thesis, University of Colorado, Boulder.

Garrison, J.L., A. Komjathy, V.U. Zavorotny, and S.J. Katzberg. 2002. Wind speed measurement using forward scattered GPS signals. IEEE Transactions on Geoscience and Remote Sensing 40(1): 50–65.

Gleason, S. 2006. Remote Sensing of Ocean, Ice and Land Surfaces Using Bistatically Scattered GNSS Signals From Low Earth Orbit. 2006. PhD Dissertation. University of Surrey, Guildford, Surrey, United Kingdom.

Huang, C.Y., Y.-H. Kuo, S.-Y. Chen, C.-T. Terng, F.-C. Chien, P.-L. Lin, M.-T. Kueh, S.-H. Chen, M.-J. Yang, C.-J. Wang, A.P. Rao. 2010. Impact of GPS radio occultation data assimilation on regional weather predictions. GPS Solutions 14(1): 35–49.

Katzberg, S.J., and J. Garrison. 1996. Utilizing GPS to Determine Ionospheric Delay Over the Ocean. NASA Technical Memorandum 4750. Washington, D.C.: NASA.

Katzberg, S.J., O. Torres, M.S. Grant, and D. Masers. 2005. Utilizing calibrated GPS reflected signals to estimate soil reflectivity and dielectric constant: results from SMEX02, Remote Sensing of Environment 100(1): 17–28.

Larson, K.M., E.E. Small, E. Gutmann, A. Bilich, P. Axelrad, and J. Braun. 2008a. Using GPS multipath to measure soil moisture fluctuations: initial results. GPS Solutions 12(3): 173–177.

Larson, K.M., E.E. Small, E.D. Gutmann, A.L. Bilich, J.J. Braun, and V.U. Zavorotny. 2008b. Use of GPS receivers as a soil moisture network for water cycle studies. Geophysical Research Letters 35(24): 1–5.

Larson, K.M., E.D. Gutmann, V.U. Zavorotny, J.J. Braun, M.W. Williams, and F.G. Nievinski, 2009. Can we measure snow depth with GPS receivers? Geophysical Research Letters 36(17): 1–5.

Larson, K.M., J.J. Braun, E.E. Small, V.U. Zavorotny, E.D. Gutmann, and A.L. Bilich. 2010. GPS multipath and its relation to near-surface soil moisture content. IEEE Journal of Selected Topics in Applied Earth Observations and Remote Sensing 3(1): 91–99.

Lowe, S.T., J.L. LaBrecque, C. Zuffada, L.J. Romans, L.E. Young, and G.A. Hajj. 2002. First spaceborne observation of an Earth-reflected GPS signal. Radio Science 37(1): 1-28.

Martin-Neira, M. 1993. A passive reflectometry and interferometry system (PARIS): application to ocean altimetry. ESA Journal 17: 331–355.

Masters, D., P. Axelrad, and S. Katzberg. 2004. Initial results of land-reflected GPS bistatic radar measurements in SMEX02, Remote Sensing of Environment 92(4): 507–520.

NRC (National Research Council). 2007. Earth Science and Applications from Space: National Imperatives for the Next Decade and Beyond. Washington, D.C.: National Academies Press.

Rivas, M.B., J. Maslanik, and P. Axelrad. 2010. Bistatic scattering of GPS signals off Arctic Sea ice. IEEE Transactions on Geoscience and Remote Sensing 48(3): 1548–1553.

Rocken, C., R. Ware, T. Van Hove, F. Solheim, C. Alber, J. Johnson, M. Bevis, and S. Businger. 1993. Sensing atmospheric water vapor with the Global Positioning System. Geophysical Research Letters 20(23): 2631–2634.

Rocken, C., J. Johnson, T. Van Hove, T. Iwabuchi. 2005. Atmospheric water vapor and geoid measurements in the open ocean with GPS. Geophysical Research Letters 32(12): 1–3.

Schreiner, W., D. Hunt, C., Rocken, and S. Sokolovskiy. 1998. Precise GPS Data Processing for the GPS/MET Radio Occultation Mission at UCAR. Pp. 103–112 in Proceedings of the 1998 National Technical Meeting of The Institute of Navigation, Long Beach, Calif., January 1998.

Small, E.E., K.M. Larson, and J. J. Braun. 2010. Sensing vegetation growth using reflected GPS signals. Geophysical Research Letters 37(L12401): 1–5.

Yunck, T.P., C.H. Liu, and R. Ware. 2000. A history of GPS sounding. Terrestrial, Atmospheric and Oceanic Science 11(1): 1–20.

Recent Progress on GNSS Seismology

LIU JINGNAN, FANG RONGXIN, and SHI CHUANG
GNSS Research Center
Wuhan University

ABSTRACT

In this paper we introduce the background of seismology development and the opportunities and challenges it faces. To verify the agreement of precise point positioning (PPP) results and seismometer results, we set up a Global Navigation Satellite System (GNSS) seismologic testing platform. The experiments reflect that the results by the two methods have great agreement. With regard to the applications of GNSS seismology, we discuss the method of using GNSS to infer epicenters of earthquakes. Eventually the perspective of GNSS seismology is put forward.

INTRODUCTION

The GNSS has been widely used to study kinematic deformation. The accuracy of GNSS kinematic positioning has been significantly improved and can reach millimeter level (Elósegui et al., 2006). At present, the receiver technology and data storage capabilities of GNSS have been greatly improved with the sampling rate up to 50 Hz (Genrich and Bock, 2006). At the same time, high-rate GNSS tracking networks have been well established, with which recent earthquakes have been successfully observed (Emore et al., 2007; Irwan et al., 2004; Langbein and Bock, 2004).

When comparing GNSS with a seismometer, the differences are remarkable (Wang et al., 2007). The digital seismometer measures the accelerations or velocities of ground motions directly (Wang et al., 2011). By contrast, GNSS estimates the position of an antenna by recording range measurements between the antenna and satellite. Moreover, the datum of GNSS positioning is satellite constellations, whereas the seismometer measures in a ground inertial reference frame.

A GNSS receiver has several advantages over a traditional seismometer. First, when the displacements of ground motions are required, GNSS can directly estimate them by range measurements and the results have no accumulated errors over time, but after integrating seismometer data to displacement, large drifts will occur. Secondly, a seismometer may be saturated in a large earthquake, in which case the instrument can not record the full amplitude of velocity or acceleration while GNSS will not be saturated in amplitude. Thirdly, a seismometer operates based on the theory of gravity, and a tilt of the instrument can bring about artificial horizontal acceleration, but a GNSS receiver will not be affected in this way. When a seismometer is saturated and unavailable to record large or nearby earthquakes, GNSS can be a feasible tool for earthquake studies.

When kinematic GNSS technology is used in seismology, it is referred to as GNSS seismology (Bock et al., 2011; Larson, 2009). In other words, using a high-rate GNSS technique to investigate the background, developing process, and explosion of earthquakes, and combining other data to determine relevant parameters can be called GNSS seismology. A high-rate GNSS receiver that is used as an instrument to capture co-seismic waves is called a GNSS seismometer.

The first experiment on GPS seismometers was reported in 1994 and was carried out by Hirahara et al. (1994) at the Disaster Prevention Research Institute of Kyoto University. In another experiment, Ge et al. (2000) fixed a GPS antenna, an accelerometer, and a velometer on a truck platform and demonstrated how GPS could recover the truck oscillation in both frequency and amplitude. They were the pioneers to reveal that GPS is able to measure large displacements with a high sampling rate.

After the 2002 Mw 7.9 Denali earthquake, Larson recovered the seismic waves with 1 Hz high-rate GPS data collected from California GPS tracking networks (Larson et al., 2003). Since then, a number of similar studies followed to use high-rate GPS technology to reconstruct waves caused by earthquakes (Bilich et al., 2008; Bock et al., 2000; Kouba, 2003; Shi et al., 2010) and to determine source parameters of earthquakes (Davis and Smalley, 2009; Ji et al., 2004; Miyazaki et al., 2004).

Kinematic high-rate GNSS positioning has been demonstrated successful for measuring seismic waves, in which relative mode is applied for GNSS data processing by assuming that one of the stations can be fixed to serve as a datum to compute absolute displacements (Bock and Prawirodirdjo, 2004; Larson et al., 2007). Obviously, a fixed datum may not be appropriate in the case of a large earthquake, and under that circumstance PPP instead of relative positioning is highly desirable. However, up to the present, there is no PPP result available to directly compare displacements from a seismometer and a GNSS receiver. The current studies of high-rate GNSS waveforms focus on the horizontal component and information on the vertical component has been ignored (Larson et al., 2007).

There are some challenges for GNSS seismology. How to improve the precision of GNSS positioning? How to realize real-time processing for huge network and ultra high-rate (20–50 Hz) GNSS data? How to explore methods of combining GNSS data and strong motion data for earthquake investigation?

For the above problems, we made some primary studies and tests as indicated in the following.

GNSS KINEMATIC POSITIONING AND TEST

Presently there are two approaches in GNSS kinematic positioning: relative positioning (RP) technique (Bock and Prawirodirdjo, 2004; Larson et al., 2007) and PPP technique (Geng et al., 2010; Kouba, 2003; Larson et al., 2003; Shi et al., 2010). In the former approach, at least one station must be fixed or tightly constrained to its known position, which will possibly experience displacement by the seismic motions. Therefore, the displacements estimated for the other stations are affected by the motions of the fixed station. In order to obtain displacements relative to a reference frame, stations not affected by the earthquake should be included as reference stations for data processing. As the position accuracy of relative positioning degrades along with the length of baselines, the inter-station distance is limited to hundreds of kilometers as in published studies (Ge et al., 2006).

In the PPP approach, where satellite clocks and orbits are fixed to pre-estimated precise parameters, for example, in the International GNSS Services (IGS) final products, the coordinates can be estimated station by station in the reference frame defined by the orbits and clocks. However, orbit and clock errors may affect the accuracy of estimated position. Hence we need to obtain high-precision orbit and clock data for positioning. In our study and test, the PPP approach was employed.

GNSS Seismometer Testing Platform

In order to compare the displacements of high-rate GNSS PPP results and seismometers results, we set up a testing platform (Figure 1). It is made of an aluminous board that is connected to the metal frame by eight springs. A GNSS antenna, seismometers recording acceleration, and an inertial measurement unit (IMU) are attached on the same board.

GNSS data were collected with a sampling rate of 50 Hz while seismometers and IMU simultaneously worked to record data of acceleration and attitude.

Data Processing Strategy

High-rate GNSS data were processed with the PANDA (position and navigation system data analyst) software, which was developed by the GNSS Research Center of Wuhan University as a multifunctional tool for GNSS research and

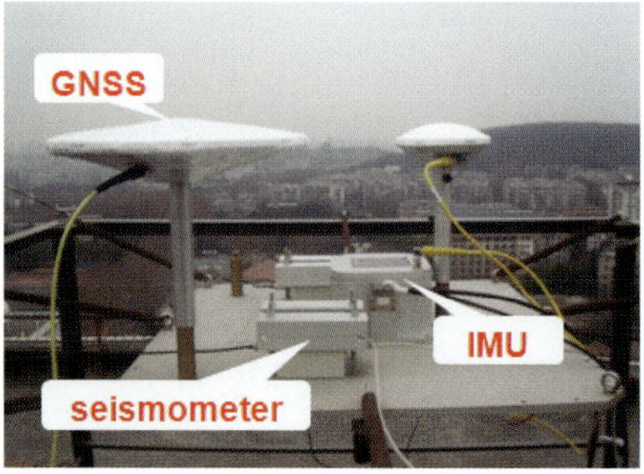

FIGURE 1 GNSS seismometer testing platform.

applications (Liu and Ge, 2003; Shi et al., 2006, 2008). Its performances in precise static and kinematic positioning were demonstrated by Ge et al. (2008) and Geng et al. (2009). In our test, PANDA software was used to compute displacements recorded by a GNSS receiver with the PPP technique. In order to correct for the tilting of the platform, IMU was used to obtain the attitude of the platform. After attitude correction, accelerometer data was integrated twice to obtain displacements.

Results and Analysis

Figure 2 shows a 50-second oscillation in the east component. The maximum displacement amplitude is about 1 cm. The amplitude of the maximum error is 5.6 mm. About 98 percent of the differences are less than 4 mm. The RMS error over the entire event is 1.8 mm.

APPLICATIONS OF GNSS SEISMOLOGY

Studies on Propagation Velocity and Earthquake Parameter Determination

Seismic waves take different times to reach various stations, and with the signals captured by GNSS we can determine the arrival times of seismic waves.

Figure 3 shows the Wenchuan earthquake seismic waveforms of the east component captured by GPS sites located at Shanghai, Wuhan, Xi'an, and Chongqing, respectively (from top to bottom).

With the time that an earthquake occurs, the times that seismic waves reach GNSS stations, and the distance between the stations and the epicenter, we can calculate the propagation speed of seismic waves. According to our calculation, the average speed of seismic wave propagation for the Wenchuan earthquake is 3.9 km/s.

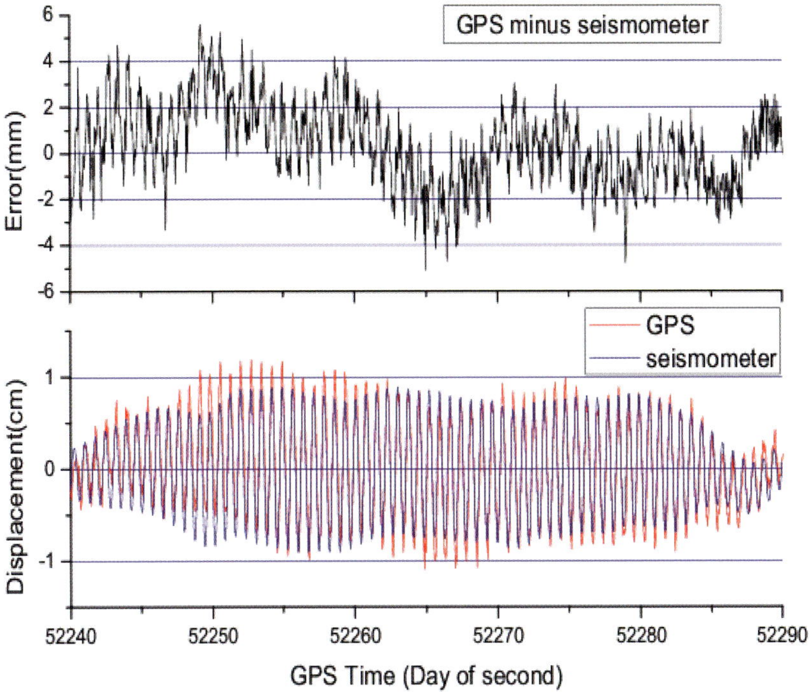

FIGURE 2 Comparison between GPS and seismometer (below), and their differences (upper).

Suppose the seismic waves propagate at equal speeds along all components, the position of the epicenter and the average speed of seismic wave propagation can be inferred with the known coordinates of the above four stations and the arrival times of seismic waves. Figure 4 describes the principle of epicenter determination with the coordinates of GNSS stations and the arrival times of seismic waves. Based on this method, GPS data of the Wenchuan earthquake are used to infer the epicenter. Compared with the result issued by the China Seismological Bureau, the inversion error is 12.5 km. The epicenter of the Chile earthquake on February 27, 2010, is also inferred with the inversion error of 27.4 km compared with the result released by the U.S. Geological Survey (USGS).

CONCLUSIONS

High-rate GNSS is going to be a new method to monitor earthquakes. With high-rate GNSS data, both seismic waveforms and the permanent offsets can be

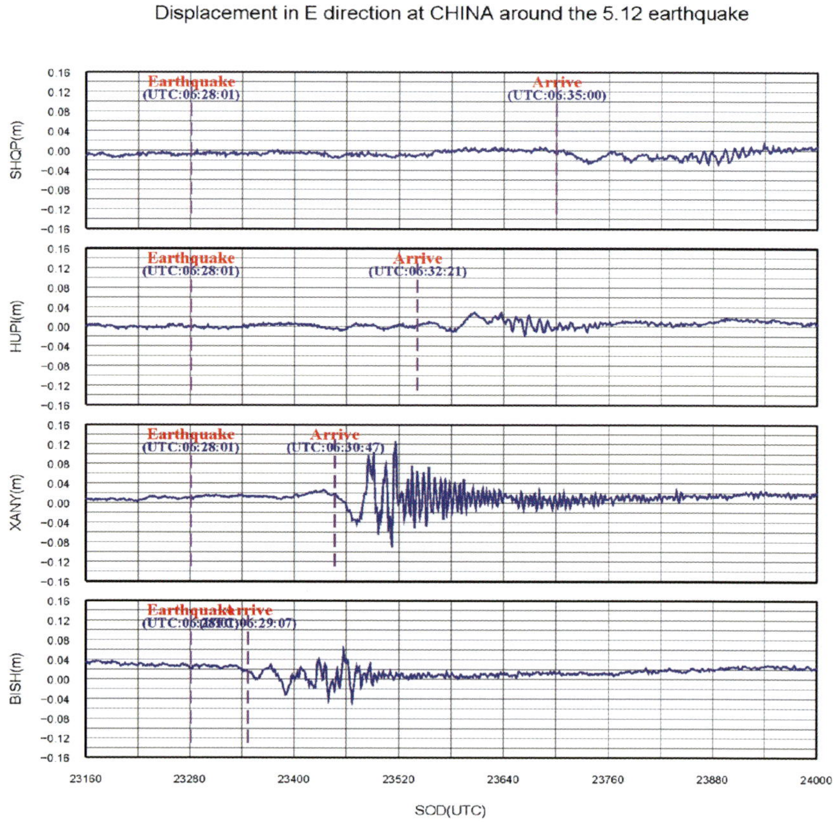

FIGURE 3 Seismic waveforms of Wenchuan earthquake in the east component captured by GPS sites located at Shanghai, Wuhan, Xi'an, and Chongqing respectively (from top to bottom).

obtained. Combining high-rate GNSS data and seismometer data, seismologic research on earthquake parameter determination and fault rupture modeling will be more scientific and reliable (Bock et al., 2011). Thus the high-rate GNSS positioning technique can be applied for real-time earthquake monitoring (Allen and Ziv, 2011), tsunami warning (Blewitt et al., 2006; Sobolev et al., 2007), and other engineering deformation monitoring.

There are still some problems with GNSS seismology. For instance, GNSS is not as sensitive as seismometers, GNSS is not applicable for small or far-field earthquakes currently, and the sampling rate is still insufficient to capture high-frequency seismic waves. These problems need to be further studied in our future research.

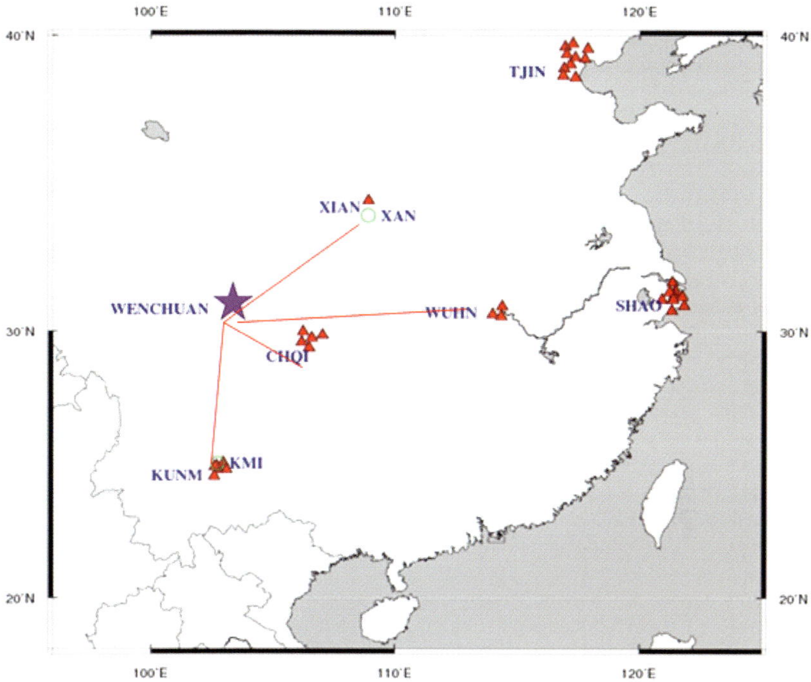

FIGURE 4 Epicenter determination with four GNSS stations.

REFERENCES

Allen, R.M., and A. Ziv. 2011. Application of real-time GPS to earthquake early warning. Geophysical Research Letters 38: L16310. doi: 10.1029/2011GL047947.

Bilich, A., J.F. Cassidy, and K.M. Larson. 2008. GPS seismology: application to the 2002 Mw 7.9 Denali Fault earthquake. Bulletin of the Seismological Society of America 98(2): 593–606.

Blewitt, G., C. Kreemer, W.C. Hammond, H.-P. Plag, S. Stein, and E. Okal. 2006. Rapid determination of earthquake magnitude using GPS for tsunami warning systems. Geophysical Research Letters 33: L11309. doi: 10. 1029/2006GL026145.

Bock, Y., and L. Prawirodirdjo. 2004. Detection of arbitrarily large dynamic ground motions with a dense high-rate GPS networks. Geophysical Research Letters 31: L06604. doi: 10.1029/2003GL019150.

Bock, Y., R.M. Nikolaidis, and P.J. De Jonge. 2000. Instantaneous geodetic positioning at medium distances with the Global Positioning System. Journal of Geophysical Research 105(B12): 28,223–28,253.

Bock, Y., D. Melgar, and B.W. Crowell. 2011. Real-time strong-motion broadband displacements from collocated GPS and accelerometers. Bulletin of the Seismological Society of America 106(6): 2904-2925. doi: 10.1785/0120110007.

Davis, J.P., and R. Smalley Jr. 2009. Love wave dispersion in central North America determined using absolute displacement seismograms from high-rate GPS. Journal of Geophysical Research 114: B11303. doi:10.1029/2009JB006288.

Elósegui, P., J.L. Davis, D. Oberlander, R. Baena, and G. Ekström. 2006. Accuracy of high-rate GPS for seismology. Geophysical Research Letters 33: L11308. doi: 10.1029/2006GL026065.

Emore, G.L., J.S. Haase, K. Choi, K.M. Larson, and A. Yamagiwa. 2007. Recovering seismic displacements through combined use of 1-Hz GPS and strong-motion accelerometers. Bulletin of the Seismological Society of America 97(2): 357–378.

Ge, L., S. Han, C. Rizos, Y. Ishikawa, M. Hoshiba, Y. Yoshida, M. Izawa, N. Hashimoto, and S. Himori. 2000. GPS seismometers with up to 20-Hz sampling rate. Earth, Planets and Space 52(10): 881–884.

Ge, M., G. Gendt, G. Dick, F.P. Zhang, and M. Rothacher. 2006. A new data processing strategy for huge GNSS global networks. Journal of Geodesy 80(4): 199–203. doi: 10.1007/s00190-006-044-x.

Ge, M., G. Gendt, M. Rothacher, C. Shi, and J. Liu. 2008. Resolution of GPS carrier-phase ambiguities in Precise Point Positioning (PPP) with daily observations. Journal of Geodesy 82(7): 389–399.

Geng, J., F.N. Teferle, C. Shi, A.H. Dodson, and J. Liu. 2009. Ambiguity resolution in precise point positioning with hourly data. GPS Solutions 13(4): 263–270. doi: 10.1007/s10291-009-0119-2.

Geng, J., F.N. Teferle, X. Meng, and A.H. Dodson. 2010. Kinematic precise point positioning at remote marine platforms. GPS Solutions 14(4): 343–350.

Genrich, J.F., and Y. Bock. 2006. Instantaneous geodetic positioning with 10–50 Hz GPS measurements: noise characteristics and implications for monitoring networks. Journal of Geophysical Research 111: B03403. doi 10.1029/2005JB003617.

Hirahara, K., et al. 1994. An Experiment for GPS Strain Seismometer. Pp. 67–76 in Japanese Symposium on GPS, December 15–16, Tokyo, Japan.

Irwan, M., F. Kimata, K. Hirahara, T. Sagiya, and A. Yamagiwa. 2004. Measuring ground deformations with 1-Hz GPS data: the 2003 Tokachi-oki earthquake (preliminary report). Earth, Planets and Space 56(3): 389–393.

Ji, C., K.M. Larson, Y. Tan, K. Hudnut, and K. Choi. 2004. Slip history of the 2003 San Simeon Earthquake constrained by combining 1-Hz GPS, strong motion, and teleseismic data. Geophysical Research Letters 31(17): L17608. doi:10.1029/2004GL020448.

Kouba, J. 2003. Measuring seismic waves induced by large earthquakes with GPS. Studia Geophysica et Geodaetica 47(4):741–755. doi:10.1023/A:1026390618355.

Langbein, J., and Y. Bock. 2004. High-rate real-time GPS network at Parkfield: utility for detecting fault slip and seismic displacements. Geophysical Research Letters 31: L15S20. doi: 10.1029/2003GL019408.

Larson, K.M. 2009. GPS seismology. Journal of Geodesy 83: 227–233. doi: 10.1007/s00190-008-0233-x.

Larson, K.M., P. Bodin, and J. Gomberg. 2003. Using 1-Hz GPS data to measure deformations caused by the Denali Fault earthquake. Science 300(5624): 1421–1424.

Larson, K.M., A. Bilich, and P. Axelrad. 2007. Improving the precision of high-rate GPS. Journal of Geophysical Research 112: B05422. doi: 10.1029/2006JB004367.

Liu, J., and M. Ge. 2003. PANDA software and its preliminary result of positioning and orbit determination. Wuhan University Journal of Natural Science 8(2B): 603–609. doi: 10.1007/BF02899825.

Miyazaki, S., K.M. Larson, K. Choi, K. Hikima, K. Koketsu, P. Bodin, J. Haase, G. Emore, and A. Yamagiwa. 2004. Modeling the rupture process of the 2003 September 25 Tokachi-Oki (Hokkaido) earthquake using 1-Hz GPS data. Geophysical Research Letters 31: L21603. doi: 10.1029/ 2004GL021457.

Shi, C., J. Geng, J. Liu, and M. Ge. 2006. Adjustment of LEO and GPS with PANDA in Precision Orbit Determination. In Proceedings of ION GNSS 2006, September 26–29, Fort Worth, Texas.

Shi, C., Q. Zhao, J. Geng, Y. Lou, M. Ge, and J. Liu. 2008. Recent Development of PANDA Software in GNSS Data Processing. In Proceedings of the Society of Photographic Instrumentation Engineers 7285, 72851S. doi: 10.1117/12.816261.

Shi, C., Y. Lou, H. Zhang, Q. Zhao, J. Geng, R. Wang, R. Fang, and J. Liu. 2010. Seismic deformation of the Mw 8.0 Wenchuan earthquake from high-rate GPS observations. Advances in Space Research 46(2): 228–235. doi:10.1016/j.asr.2010.03.006.

Sobolev, S.V., A.Y. Babeyko, R. Wang, A. Hoechner, and R. Galas. 2007. Tsunami early warning using GPS-shield arrays. Journal of Geophysical Research 112: B08415. doi: 10.1029/2006JB004640.

Wang, G., D.M. Boore, G. Tang, and X. Zhou. 2007. Comparisons of ground motions from colocated and closely spaced one-sample-per-second Global Positioning System and accelerograph recordings of the 2003 M6.5 San Simeon, California earthquake in the Parkfield Region. Bulletin of the Seismological Society of America 97(1B): 76–90. doi: 10.1785/0120060053.

Wang, R., B. Schurr, C. Milkereit, Z. Shao, and M. Jin. 2011. An improved automatic scheme for empirical baseline correction of digital strong-motion records. Bulletin of the Seismological Society of America 101(5): 2029–2044. doi: 10.1785/0120110039.

Precision Agriculture: Opportunities and Challenges

MICHAEL O'CONNOR
O'C and Associates

INTRODUCTION TO PRECISION AGRICULTURE

Precision Agriculture (or Precision Farming) is a term used to describe the use of technology to better measure and control crop production on a site-specific basis to improve efficiency. Such improvements include:

- More efficient application of inputs (seed, fertilizer)
- More effective utilization of tillage equipment
- Improved crop and field measurements
- Better farm management decisions

While computers and electronics have been used in crop production since the 1970s, GNSS has been a key enabling technology for Precision Agriculture beginning in the mid-1990s.

THE PRECISION FARMING FEEDBACK LOOP

Historically, the process of crop production has been an "open loop" process, with only qualitative or imprecise feedback methods available to growers. This process is illustrated in Figure 1.

Growers generally use the best information available to them, including the crop history on their farm (e.g., for crop rotation); information about available seed types; the current costs of inputs such as fertilizer, seed, fuel, and labor; climate history for their area; and the recent weather for their area. This information

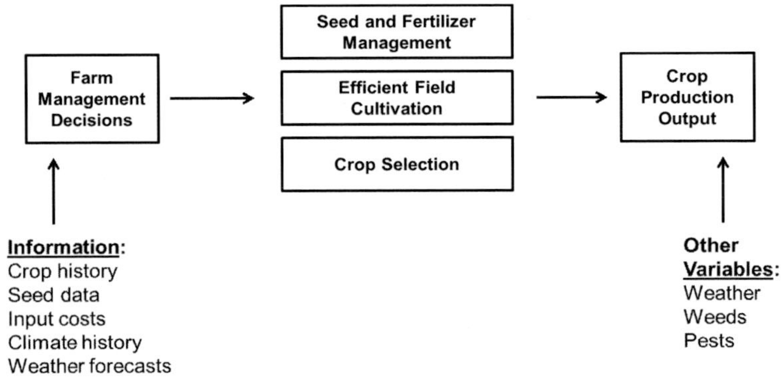

FIGURE 1 "Open loop" crop production.

is used to make fundamental farm management decisions such as which crop to plant in each field, which seeds to use, when to plant, how much fertilizer to use, how to till the ground, and what planting pattern and spacing to use in the field. These decisions are made with the goal of optimizing the farm's operation and maximizing crop production output for the farm. Unfortunately, there are also several dominant external factors that affect crop production—in particular the weather and weed and pest infestations.

With the introduction of Precision Agriculture, including advancements in electronics, computers, software, and sensors, growers now have better tools to manage their crop production. These tools are shown in blue in Figure 2 and are described in more detail below.

Yield Measurements

The practice of using moisture and grain flow sensors in combine harvesters to measure yield was the first precision agriculture practice to become widely adopted. While the practice of measuring yield on-the-go was introduced in the 1980s, the integration of yield measurements with GNSS in 1994 was a revolution. GNSS-based yield monitors gave farmers a tool to collect site-specific information about their crop production and to generate maps showing in-field yield variability on their farms. Today nearly every combine harvester manufactured and sold in North America includes a yield monitor.

Soil Nutrient Measurements

Nitrogen, phosphorous, potassium, and other soil nutrients are critical to plant health. For thousands of years growers have been aware of the importance of soil

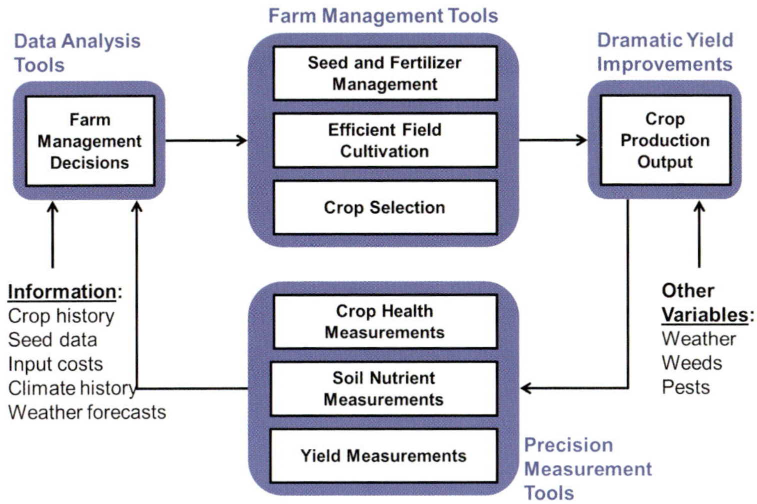

FIGURE 2 "Closed loop" crop production using Precision Agriculture practices.

nutrients and have used organic fertilizers to amend their soils before planting. In more recent times, scientists and agronomists have learned more about the specific chemical needs of crops and have developed specialized fertilizers to directly target those needs.

Today, growers have access to services that will collect soil samples from their fields, mail those samples to a chemistry lab, and provide a map showing the site-specific nutrient levels within the growers' fields. Unfortunately, this process is slow, highly seasonal, and labor intensive.

Crop Health Measurements

Growers are now able to better measure the health of their crops during the season. Field scouting techniques using GNSS are becoming popular in North America, and growers are beginning to utilize remote imagery from satellites or aircraft with multispectral imaging cameras. Near-real-time sensing of crop health can drive in-season management decisions such as pesticide application and in-season nutrient management (Figure 3).

Crop Selection

While not directly related to computers, electronics, or sensors, crop selection is becoming one of the most important variables for a grower to manage. Seed selection has a dramatic impact on input costs as well as yield. Because they are

FIGURE 3 Crop health measurement techniques.

so effective, genetically modified seeds have become widely adopted in the United States. These seeds are engineered for characteristics such as higher yields, pest resistance, and herbicide resistance. These results are highly visible to the grower, and the economic value is compelling.

Efficient Field Cultivation

In addition to allowing for site-specific measurements within a field, GNSS has also enabled robotic automation of farm equipment. Products introduced to the market in the year 2000 enable tractors, sprayers, and harvesters to steer through a field, hands-free, with sub-inch accuracy.

Automated steering provides clear benefits to growers (Figure 4). It allows equipment to run around the clock—regardless of visibility—in the daytime, nighttime, or in the fog. Precision-steered vehicles experience 8 to 10 percent less overlap between passes than human-steered vehicles, which leads to lower fuel, labor, and input costs. Also, less overlap results in more rows in the field, which leads to greater yields. The results of hands-free steering are visible, and the economic value is compelling.

Seed and Fertilizer Management

In addition to controlling the steering of farm vehicles, GNSS enables solutions that can control the application of field inputs on a site-specific basis. Powerful tools are now available to growers that allow real-time adjustment of seed and fertilizer rates. These rates are established by software "prescriptions," which are created based on a variety of data, including the yield monitoring and soil nutrient data described above.

Adoption of these techniques has been relatively slow compared to genetically modified seeds and automated steering, primarily because the soil nutrient

FIGURE 4 Example of the benefits of hand-free precision steering.

measurement process is slow and tedious, and the results of performing these variable rate application techniques are difficult to measure because of the influence of so many other variables on crop output (such as weather).

TODAY'S LEADING CHALLENGES FOR PRECISION AGRICULTURE

Adoption rates of seed genetics and precision steering have exceeded 50 percent in several geographic markets because of their visible and compelling value. However, grower adoption of seed and fertilizer management continues to lag.

More efficient use of field inputs—particularly nitrogen fertilizer—is essential for several reasons. For one, fertilizer costs are rising. Fertilizer sales now exceed $18 billion annually in the United States and represent between 30–50 percent of the cost of production for wheat and corn on most farms. In addition, worldwide fertilizer use is on the rise. Globally, the rate of nitrogen use is outpacing increases in population and in arable land. Oxygen depletion triggered by excessive nitrogen and phosphorous levels, primarily caused by fertilizer runoff, is becoming a serious problem in several major waterways. The U.S. National Academy of Engineering has listed "Managing the Nitrogen Cycle" as one of its 14 grand engineering challenges for the 21st century.

It is the author's belief that adoption of Precision Agriculture for seed and fertilizer management will improve when three key challenges have been overcome:

- Improving GNSS signal availability
- Improving the efficiency of soil measurements
- Analyzing data across multiple farms

Improving GNSS Signal Availability

GPS alone does not provide sufficient field coverage in many farm environments. Hilly or mountainous terrain is rare, but tree lines are a common issue, as shown in Figure 5. Reliability of the signal has been a barrier to adoption in areas such as the southwestern United States. The majority of growers in the United States 10 years ago were unfamiliar with GPS. Today many growers are educated enough in GNSS to ask about signal availability and reacquisition.

Most high-precision systems sold in North America now offer GLONASS capability to augment GPS for signal availability. As more satellite signals-in-space become available, availability will continue to improve in these difficult environments, and more growers will view GNSS as a reliable solution for their needs.

Improving the Efficiency of Soil Measurements

As described above, current soil nutrient measurement techniques are slow, expensive, and inaccurate. To reduce costs, many growers employ "zone" sampling techniques, in which one to five samples are collected across an entire field (typically 80 to 160 acres) based on soil texture zones. More progressive growers who practice "high density" sampling typically collect samples on a grid at one sample per 2.5 acres. Studies show that, given the spatial decorrelation of soil nutrients, sampling to at least one sample per acre is required to accurately interpolate nutrient levels across a field.

Unfortunately, the financial cost of applying extra fertilizer in a field is significantly lower than the potential yield reduction caused by an under-application of fertilizer. Until growers can measure nutrients affordably and at higher density in near-real-time (particularly nitrogen, which is water soluble and highly time-dependent), they will continue to over-apply fertilizers to ensure high yields.

FIGURE 5 Farm fields lined with GNSS-barriers.

Data Analysis Across Multiple Farms

As described above, many variables affect crop yield. These include rainfall, temperature, humidity, wind, soil type, and tillage practices, just to name a few.

Unfortunately, today, farms are "islands" of information. For a single farm to collect enough information to measure the value created by a new farming practice or a particular seed hybrid would take many years. A single farm cannot produce enough data to provide meaningful statistical significance.

Dramatic improvement in data analysis will be possible when information can be correlated across many farms. The wireless data connections and software tools to enable such analysis are just coming into practice now on farms. Once these practices become more widespread, the value of Precision Farming will become clearer to growers and adoption of these practices will increase.

SUMMARY

Precision Agriculture is enabling more efficient application of inputs (seed, fertilizer), more effective utilization of tillage equipment, improved crop and field measurements, and better farm management decisions. With advancements in electronics, computers, software, and sensors, growers now have better tools to manage their crop production.

In the United States, adoption rates have been very fast for some of these technologies, such as seed genetics and precision steering, because of their high visibility and compelling value. However, grower adoption of seed and fertilizer management continues to lag. This is a serious problem because improper use of fertilizer is economically wasteful for growers and is also causing harm to waterways and underground water sources.

It is the author's belief that adoption of Precision Agriculture for seed and fertilizer management will improve when three key challenges have been overcome: (1) improving GNSS signal availability, (2) improving the efficiency of soil measurements, and (3) when growers take advantage of the ability to perform data analysis across multiple farms.

Integrity Lessons from the WAAS Integrity Performance Panel

TODD WALTER AND PER ENGE
Stanford University

BRUCE DeCLEENE
Federal Aviation Administration

ABSTRACT

The Wide Area Augmentation System (WAAS), a system to allow the use of the Global Positioning System (GPS) for many phases of flight within the United States, was implemented because its benefits are significant. It provides guidance throughout the national airspace. It supports approaches with vertical guidance to more runway ends in the United States than all other landing systems combined. It does this without requiring local navigational aids. It also can support procedures such as curved approaches and departures. These and other benefits motivated the effort to create and certify this new system. However, WAAS is unlike previous navigation systems fielded by the Federal Aviation Administration (FAA). Historically, the FAA has implemented relatively simple and distributed systems. Each only affects a small portion of the airspace and each is maintained independently of the others. WAAS, in contrast, is a complex and centralized system that provides guidance to the whole airspace. Consequently, the certification for WAAS proceeded very cautiously.

A unique aspect of WAAS compared to traditional terrestrial navigational aids is that it is inherently a non-stationary system. It relies on satellites that are constantly in motion and that may change their characteristics. Additionally, the propagation of the satellite signals varies with local ionospheric and tropospheric conditions. Thus, the system has differing properties over time and space. However, the system requirements apply to each individual approach. In particular, the integrity requirement, that the confidence bound fails to contain the true error in fewer than 1 in 10 million approaches, must apply to all users under all foreseeable operational conditions. To ensure that the integrity requirement would be met, the FAA formed the WAAS Integrity Performance Panel (WIPP). The role of the WIPP is to independently assess the safety of WAAS and to recommend system improvements. To accomplish these tasks for initial certification, the WIPP had to determine how to interpret the integrity requirement for WAAS, develop algorithms to meet this requirement, and ultimately validate them.

INTRODUCTION

WAAS monitors the GPS and provides both differential corrections to improve the accuracy and associated confidence bounds to ensure the integrity. WAAS utilizes a network of precisely surveyed reference receivers located throughout the United States. The information gathered from these WAAS Reference Stations (WRSs) monitors GPS and its propagation environment in real-time. However, the WAAS designers had to be aware of the limitations of its monitoring. The measurements that it makes are corrupted by noise and biases causing certain fault modes to be difficult to detect. Because it is a safety-of-life system, WAAS must place rigorous bounds on the probability that it is in error, even under faulted conditions.

In late 1999, concerns arose over the original WAAS design and the process by which WAAS was to be proven safe. In response, the FAA created the WIPP. The WIPP is a body of GPS and system safety experts chartered to assess the system engineering and safety design of WAAS and recommend required changes. The WIPP consists of members from government (FAA, Jet Propulsion Laboratory), industry (Raytheon, Zeta, MITRE), and academia (Stanford University). They first convened in early 2000 to address the integrity and certification of WAAS.

Primarily WIPP quantified the degree to which WAAS mitigated the system vulnerabilities. Over its first two years, WIPP changed the design of several system components where the system could not satisfactorily demonstrate the required level of integrity. As each threat was addressed, WIPP built upon what it had learned.

Some of the main lessons that emerged from WIPP are:

- The aviation integrity requirement of 10^{-7} per approach applies in principle to each and every approach. It is not an ensemble average over all conditions.
- Validated threat models are essential both to describe what the system protects against and to quantitatively assess how effectively it provides such protection.
- The system design must be shown to be safe against all fault modes including external threats, addressing the potential for latent faults just beneath the system's ability to detect them. This approach is unlike

conventional non-aviation differential systems that presume no failures exist unless consistency checks fail.

- The safety analysis must protect all allowed geometries. It does not protect just the all-in-view case; all subset combinations that support the operation must be safe as well.
- The small numbers associated with integrity analysis are not intuitive. Careful analysis must take priority over anecdotal evidence.

These lessons will be described in greater detail. Of these lessons, the need for threat models is the most important and originally was the most lacking. Threat models describe events or conditions that may cause harm to the user. In this case, harm is referred to as hazardously misleading information (HMI). It is defined as a true error that is larger than the guaranteed protection level (PL). WAAS provides differential corrections that are applied to the received pseudoranges from GPS. At the same time, confidence bounds are also supplied to the user. These bounds are combined with the geometry of satellites visible to the user to calculate the PL. In order to use the calculated position for navigation, the PL must be small enough to support the operation. The user only has real-time access to the PL, not the true error. Thus, HMI arises if the user has been told that the error in position is small enough to support the operation, but in fact, it is not.

The threat models must describe all known conditions that could cause the true errors to exceed the predicted confidence bounds. Having a comprehensive list is essential to achieving the required level of safety and it also drives the system design. Additionally, restricting the scope of the threats is necessary for practical reasons. It is not possible to create a system that can protect against every conceivable threat. Fortunately, many such threats are either unphysical or extremely improbable. Restricting threats to those that are sufficiently likely is necessary for creating a practical system.

INTEGRITY REQUIREMENT

The integrity requirement for precision approach guidance (down to 200′ above the ground) is 1-2 $\times$ 10^{-7} per approach (ICAO, 2006). There is a general understanding that this probabilistic requirement applies individually to every approach. This definition is further refined in the WAAS specification (2: FAA-E-2892C WAAS Specification) as applying at every location and time in the service volume. It is not acceptable for one airport to have less integrity simply because a different aircraft hundreds of miles away has margin against the requirement. Similarly, with the non-stationary characteristics arising from effects such as the orbiting satellites, it is not appropriate for operations to continue during an interval when the integrity requirement is not met, just because it is exceeded for the rest of the day. Generally, this can be restated as meaning that the probability of hazardously misleading information (HMI) must be at or below

1×10^{-7} for an approach at the worst time and location in the service volume for which the service is claimed to be available. Despite this apparent understanding, a more detailed discussion of the interpretation is instructive.

The integrity requirement is that the positioning error (PE) must be no greater than the confidence bound, known as the PL, beyond the specified probability. Confusion may result because the requirement is probabilistic, yet at the worst time and place, the errors appear deterministic. Instead, the requirement should be viewed as applying to a hypothetical collection of users under essentially identical conditions. The collection of users, referred to as the ensemble, must be hypothetical in this case because satellite navigation systems and their associated errors are inherently non-stationary. Any true ensemble would average over too many different conditions, combining users with high and low risk. Thus, we must imagine an ensemble of users, for each point in space and time, whose errors follow probability distributions specific to that point.

Of course, there can only be one actual user at a given point in space and time. That user will experience a specific set of errors that combine to create the position error. These errors are comprised of both deterministic and stochastic components. The distinction is that if we could replicate the conditions and environment for the user, the deterministic components would be completely repeatable. Thus, these errors would be common mode; all users in our ensemble would suffer them to the same degree. On the other hand, stochastic errors such as thermal noise would differ for each user in our ensemble. Overall, these components combine to form a range of possible errors whose magnitudes have differing probability. When we look at a very large number (approaching infinity) of hypothetical users in the ensemble, some will have errors that exceed the protection level while most will not. The fraction of users that exceed the PL can be used to determine the probability of an integrity failure under those conditions.

The difficult aspect of applying this philosophy is defining equivalent user conditions and then determining the error distributions. A circular definition is that user conditions can be called equivalent if they carry the same level of risk. A more practical approach is to exploit prior knowledge of the error sources. For example, if it were known that an error source only has a definite temperature dependency, then the ensembles should be formed over all users in narrow temperature ranges. The error distributions and probability of exceeding the PL would be calculated for each ensemble, and the integrity requirement would have to be met for the most difficult case for which availability is claimed. Unfortunately, true error sources usually have multiple dependencies, and these dependencies are different between the various error sources. Thus, the ensembles may need to be formed over narrow ranges of numerous parameters. However, great care must be taken because, if certain dependencies are not properly recognized, the ensembles may unknowingly average over different risk levels.

The restatement of the requirement that it applies to the worst time and location is misleading because it is acceptable to average against certain conditions. Some events may be sufficiently rare to ignore altogether. If, under similar conditions, the *a priori* likelihood is well below 1×10^{-7} per approach (considering exposure time to the failure), then there may not be any need to provide additional protection. The worst time and place should not be viewed as when and where this unlikely event occurs. The event need only be considered if it is sufficiently likely to occur, if when and where it is most likely to occur can be predicted ahead of time, or if it is strongly correlated with an observable. Even if the event is not sufficiently rare to be ignored, its *a priori* probability may be utilized provided the event remains unpredictable and immeasurable. Thus, the conditions where the event is present may be averaged with otherwise similar conditions without the event. Taking advantage of such *a priori* probabilities must be approached very cautiously on a case-by-case basis.

The goal is to ensure that all users are exposed to risk at no greater than the specified rate of 10^{-7} per approach. Thus, ensembles that cannot be correlated in some way with user behavior or an observable parameter do not make sense. For example, users may tend to fly to the same airport at the same time of day or during a certain season. Therefore, an ensemble of all users with a specific geometry at a certain location and certain time of day, but theoretically infinitely extended forwards and backwards over adjacent days, is reasonable. On the other hand, an ensemble of all users whose thermal noise consists of five-sigma errors aligned in the worst possible direction is neither realistic nor practical. The latter example attempts to combine rare and random events into a unifying ensemble that cannot be made to correspond to user behavior or to any practically measurable quantity. In general, conditions leading to high risk that are both rare and random can be averaged with lower risk conditions. The requirement for rarity seeks to assure that users do not receive multiple exposures to the high-risk condition, while the requirement for randomness seeks to avoid a predictable violation of the integrity requirement. Correlation with conceivable user behavior must be a determining factor when deciding whether or not to average the risk. Similarly, a correlation with a system observable should be exploited to protect the user when performance goes out of tolerance.

Deciding how to define the ensembles provides the necessary information for determining the error distributions. Components will largely be divided into noise-like contributions, with some spread in their values, and bias-like contributions whose values are seen as fixed although unknown. Although many of these error sources may be deterministic, practically they may need to be described in stochastic terms. Many error sources fall into this category, including ionosphere, troposphere, and multipath. If we knew enough about the surrounding environments, we could predict their effects for each user. However, because it is usually not practical to obtain this information, it may be acceptable to view these effects as unpredictable as long as their effects cannot be correlated with user behavior.

Knowledge of the error characteristics is very important in evaluating system design. While impossible to know fully, many important characteristics such as dependencies may be recognized. This knowledge allows proper determination of the error distributions. After defining the individual distributions, the correlations between them must be established. Many deterministic error sources may affect multiple ranging sources simultaneously. Correlated deterministic errors may add together coherently for a specific user. Such effects require larger increases in the protection level than if the errors were uncorrelated. If these effects are not recognized and treated appropriately, the integrity requirement will not be met and the user will suffer excessive risk. Although the form of the protection level equations given in ICAO (2006) and FAA (2009) suggest that all error sources are independent, zero-mean, and gaussian, this is not the case under all operating conditions. Each error source must be carefully analyzed, both individually and in relation to the other sources. Only then can the appropriate confidence bounds be determined.

ERROR MODELING

Each individual error source has some probability distribution associated with it. This distribution describes the likelihood of encountering a certain error value. Ideally, smaller errors are more likely than larger errors. Generally, this is true for most error sources. The central region of most error sources can be well described by a gaussian distribution. That is, most errors are clustered about a mean (usually near zero) and the likelihood of being farther away from the mean falls off according to the well-known model. This is often a consequence of the central-limit-theorem that states that distributions tend to approach gaussian as more independent random variables are combined.

Unfortunately, the tails of the observed distributions rarely look gaussian. Two competing effects tend to modify their behavior. The first is clipping. Because there are many cross-comparisons and reasonability checks, the larger errors tend to be removed. Thus, for a truly gaussian process, outlier removal would lead to fewer large errors than would otherwise be expected. The second effect is mixing. The error sources are rarely stationary. Thus, some of the time the error might be gaussian with a certain mean and sigma and at other times have a different distribution. Such mixing may result from a change in the nominal conditions or from the introduction of a fault mode. Mixing generally leads to broader tails or large errors being more likely than otherwise expected.

The mixing causes additional problems. If the error processes were stationary, it would be possible to collect as large a data set as practical and then conservatively extrapolate the tail behavior using a gaussian or other model. However, because the distribution changes over time, it is more difficult to predict future performance based on past behavior. Furthermore, mixing leads to more complicated distributions whose tails are more difficult to extrapolate. With enough mixing, it can be very difficult to characterize the underlying distributions at all. Figure 1 is an example of a mixed distribution. The majority

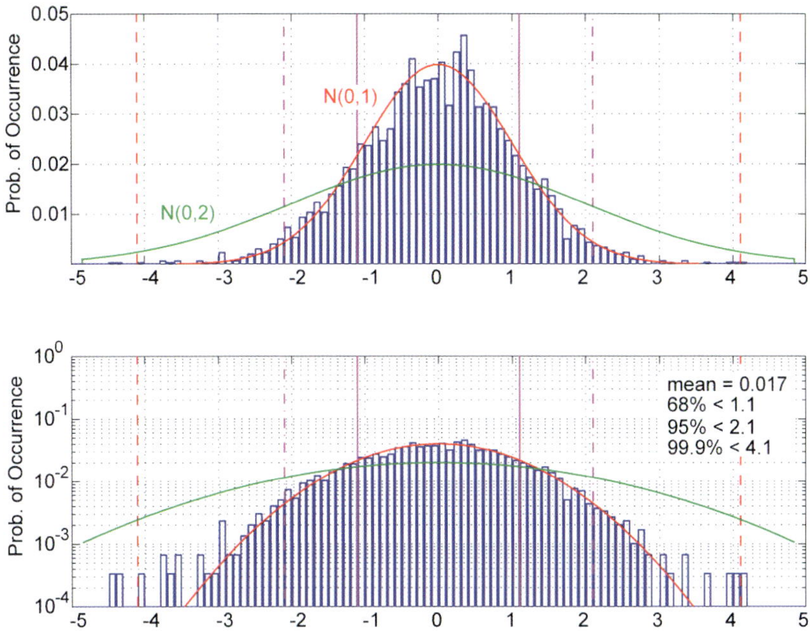

FIGURE 1 Simulated probability distribution composed of a mixed gaussian: 2,900 points with unity variance and 100 points with a variance of four. The top and bottom graphs are the same data displayed on a linear scale (top) and log-scale (bottom).

of the data points are selected from a zero-mean gaussian with unity variance. A few of the points are selected from a zero-mean gaussian with a variance of four. This plot contains some very typical features of the real data we collect. The majority of the data conforms very well to a gaussian model, while the tails usually contain outliers. Sampling issues are usually significant because it is very difficult to obtain large amounts of independent data. Thus, just by looking at the graph it is difficult to determine the actual distribution.

The central-limit-theorem causes error distributions to approach gaussian as several independent sources are combined. Certainly, the main body of collected data tends to be gaussian in appearance. The tails are more difficult to discern. A generalized mixed gaussian description is appropriate. Here, the errors can be described as gaussian where the mean and variance are also drawn from some joint probability distribution.

$$\varepsilon \in N(\mu,\sigma)$$

$$\mu,\sigma \in pr(\mathrm{M},\Sigma)$$

At any given instant, the error is gaussian, but its mean and variance have some uncertainty. By understanding the extent of the possible means and variances we can overbound the worst-case. Additional information ideally allows us to partition the space and distinguish when larger bounds are needed versus when smaller ones can be provided.

Nominally, we expect the distribution to be zero-mean and have some well-defined variance. Some small fraction of the time the error may still be zero-mean, but have a larger variance as depicted in Figure 1. During a fault mode the mean may grow in magnitude, but the variance may stay roughly the same as nominal (of course other variances are possible). Restricting the error distribution to this class distribution allows the analysis to become tractable.

Of course, it is impossible to truly know the real distribution, particularly to 10^{-7} confidence. The use of a model like this must be accepted by a body of experts such as the WIPP who can assert that it is valid based on physical knowledge of the system design, supporting data, and simulation. This combination is essential for describing the tail behavior. A physical understanding of the error process is essential to describing expected behavior. Data must be collected in sufficient quantity and under many conditions. The physical knowledge must be exploited to determine what the worst-case conditions are and how data should be reduced. For example, severe ionospheric behavior is correlated with solar events and magnetic disturbances. Data must be collected during some of the most extreme operating conditions. Finally, simulation may be used to confirm that the models constructed are consistent with the observations.

Physical knowledge of the system is essential. Any information on the physical processes behind the error source can be used to separate mixtures and create better-defined distributions. For example, multipath can be related to the surrounding environment. Large reflections tend to occur at lower elevation angles. Partitioning data by elevation angles may reduce mixing. Changes to multipath can be related to changes in satellite position and to changes in the environment. Excessive multipath can sometimes be related to specific reflectors. Additionally, the magnitude of multipath errors can be bounded by limiting the number of reflectors and strength of the reflected signals.

Data are also essential. The data must be sufficient to support assumptions or validate system performance to the degree to which the safety of the system relies on that data. It is not sufficient to collect a day or two of randomly selected data; data collected over many days under extreme conditions are needed. Examples include tropospheric data from many different climates, ionospheric data from the worst times in the 11-year solar cycle, multipath data from the most cluttered environments, etc. Rare events are unlikely to be captured in small data sets. Large data sets, taken over long time periods, are more likely to capture postulated events. Having data containing these events provides better insight into their effect.

THREAT MODELS

Threat models describe the anticipated events that the system must protect the user against and conditions during which it must provide reliably safe confidence bounds. Each threat model must describe the specific nature of the threat, its magnitude, and its likelihood. Together, the various threat models must be comprehensive in describing all reasonable conditions under which the system might have difficulty protecting the user. Ultimately they form a major part of the basis for determining if the system design meets its integrity requirement. Each individual threat must be fully mitigated to within its allocation. Only when it can be shown that each threat has been sufficiently addressed can the system be deemed safe.

WAAS was developed primarily to address existing threats to GPS. However, it also runs the risk of introducing threats in absence of any GPS fault. By necessity, it is a complex system of hardware and software. Included in any threat model must be self-induced errors. Some of these errors are universal to any design while others are specific to the implementation. For example, the software design assurance of WAAS reference receivers was based on market availability of equipment, so reference receivers' software faults are a unique threat that has to be mitigated through system integrity monitoring. The following is a high-level list of generic threats. While it is not comprehensive, it does include the most significant categories either for magnitude of effect or likelihood. There are numerous other threats that have a smaller effect, are less likely, or are implementation specific.

High-Level Threat List

- Satellite
  - Clock/ephemeris error
  - Signal deformation
  - Code carrier incoherency
- Ionosphere
  - Local non-planar behavior
    - Well-sampled
    - Undersampled
- Troposphere
- Reference receiver
  - Multipath
  - Thermal noise
  - Antenna bias
  - Survey errors
  - Receiver errors
- Master station
  - Space vehicle (SV) clock/ephemeris estimation errors
  - Ionospheric estimation errors

— SV Tgd estimation errors
— Receiver IFB estimation errors
— WRS clock estimation errors
— Communication errors
— Broadcast errors

- User errors

The following sections provide greater detail for each threat, although the true details depend on implementation and must be decided by the service provider.

SV Clock/Ephemeris Estimation Errors

Satellites suffer from nominal ephemeris and clock errors even when there are no faults in the GPS system (Creel et al., 2007; Heng et al., 2011; Jefferson and Bar-Sever, 2000; Warren and Raquet, 2002). Additionally, the broadcast GPS clock and ephemeris information may contain significant errors in the event of a GPS system fault or erroneous upload. Such faults may create jumps, ramps, or higher-order errors in the GPS clock, ephemeris, or both (Gratton et al., 2007; Hansen et al., 1998; Heng et al., 2010; Rivers, 2000; Shank and Lavrakas, 1993). Such faults may be created by changes in state of the satellite orbit or clock or simply due to the broadcasting of erroneous information. Either the user or the system may also experience incorrectly decoded ephemeris information.

The user differential range error (UDRE), a term designed to describe residual satellite errors, must be sufficient to overbound the residual errors in the corrected satellite clock and ephemeris.

Signal Deformations

The International Civil Aviation Organization (ICAO) has adopted a threat model to describe the possible signal distortions that may occur on the GPS L1 CA code (ICAO, 2006). These distortions will lead to biases that depend upon the correlator spacing and bandwidth of the observing receivers. Such biases would be transparent to a network of identically configured receivers (Hsu et al., 2008; Phelts et al., 2009; Wong et al., 2010).

The UDRE must be sufficient to overbound unobservable errors caused by signal deformation. Unobservable errors are those that cannot be detected to the required level of integrity.

Code-Carrier Incoherency

Another threat is that a satellite may fail to maintain the coherency between the broadcast code and carrier. This fault mode occurs on the satellite and is unrelated to incoherence caused by the ionosphere. This threat causes either a step or a rate of change between the code and carrier broadcast from the satellite. This threat has never been observed on the GPS L1 signals, but it has been observed on WAAS geostationary signals and on the GPS L5 signal (Gordon et al., 2010; Montenbruck et al., 2010).

The UDRE must be sufficient to overbound unobservable errors caused by incoherency. Unobservable errors are those that cannot be detected to the required level of integrity.

Ionosphere and Ionospheric Estimation Errors

The majority of the time, mid-latitude ionosphere is easily estimated and bounded using a simple local planar fit. However, periods of disturbance occasionally occur where simple confidence bounds fall significantly short of bounding the true error (Walter et al., 2001). Additionally, in other regions of the world, in particular equatorial regions, the ionosphere often cannot be adequately described by this simple model (Klobuchar et al., 2002; Rajagopal et al., 2004). Some of these disturbances can occur over very short baselines, causing them to be difficult to describe even with higher-order models. Gradients larger than 3 m of vertical delay over a 10 km baseline have been observed, even at midlatitude (Datta-Barua et al., 2002, 2010).

The broadcast ionospheric grid format specified in the Minimum Operational Performance Standards (MOPS)[1] also limits accuracy and integrity. The simple two-dimensional model and assumed obliquity factor may not always provide an accurate conversion between slant and vertical ionosphere. There will also be instances where the five-degree grid is too coarse to adequately describe the structure of the surrounding ionosphere.

There are times and locations where the ionosphere is very difficult to model. This problem may be compounded by poor observability (Sparks et al., 2001; Walter et al., 2004). Ionospheric Pierce Point (IPP) placement may be such that it fails to sample important ionospheric structures. This may result from the intrinsic layout of the reference stations and satellites, or from data loss through station, satellite, or communication outages. As a result, certain ionospheric features that invalidate the assumed model can escape detection.

Finally, because the ionosphere is not a static medium, there may be large temporal gradients in addition to spatial gradients. Rates of change as large as four vertical meters per minute have been observed at mid latitudes (Datta-Barua et al., 2002).

The grid ionospheric vertical error (GIVE), a term designed to describe residual ionospheric errors, must account for inadequacies of the assumed ionospheric model, restrictions of the grid, and limitations of observability. The GIVE must be sufficient to protect against the worst possible ionospheric disturbance

[1] WAAS Minimum Operational Performance Specification (MOPS), RTCA document DO-229D.

that may be present in that region given the IPP distribution. Additionally, since each ionospheric correction does not time out until after 10 minutes, the GIVE and the old but active data (OBAD) terms must protect against any changes in the ionosphere that can occur over that time scale.[2] Because the physics of the ionosphere are incompletely understood, the most practical ionospheric threat models are heavily data driven and contain a large amount of conservatism.

Tropospheric Errors

Tropospheric errors are typically small compared to ionospheric errors or satellite faults. Historical observations were used to formulate a model and analyze deviations from that model (Collins and Langley, 1998). A very conservative bound was applied to the distribution of those deviations. The model and bound are described in the MOPS and Standard and Recommended Procedures (SARPS) (ICAO, 2006).[3] These errors may affect the user both directly through their local troposphere and indirectly through errors at the reference stations that may propagate into satellite clock and ephemeris estimates. The user protects against the direct effect using the specified formulas.

The master station must ensure that the UDRE adequately protects against the propagated tropospheric errors and their effect on satellite clock and ephemeris estimates. Of particular concern are the statistical properties of these error sources. These errors may be correlated for long periods and will produce correlated errors across all satellites at a reference station and each receiver at the reference station.

Multipath and Thermal Noise

Multipath is the most significant measurement error source. It limits the ability to estimate the satellite and ionospheric errors. It depends upon the environment surrounding the antenna and the satellite trajectories. While many receiver tracking techniques can limit its magnitude, its period can be 10 minutes or greater (Shallberg and Sheng, 2008; Shallberg et al., 2001). Additionally, it contains a periodic component that repeats over a sidereal day. Thus, severe multipath may be seen repeatedly for several days or longer.

Because all measurements that form the corrections and the UDREs and GIVEs are affected by multipath, great care must be used to bound not only its maximum extent but also its other statistical characteristics (e.g., non-gaussian, non-white, periodic). There is potential for correlation between measurements and between antennas at a single reference site. Additionally the local environment may change either due to meteorological conditions (snow, rain, ice) or physical changes (new objects or structures placed nearby).

---

[2] Ibid.

[3] Ibid.

If carrier smoothing is used to mitigate multipath, then robust cycle slip detection is essential. Half integer cycle slips have been observed on many different types of receivers. In one case, several half cycle slips were observed in the same direction each several minutes apart resulting in a several meter error. Cycle slip detection must be able to reliably catch unfortunate combinations of L1 and L2 half and full integer cycle slips in order to achieve an unbiased result.

Antenna Bias

Look-angle dependent biases in the code phase on both L1 and L2 are present on reference station and GPS satellite antennas (Haines et al., 2005; Shallberg and Grabowski, 2002). These biases may be several tens of centimeters. In the case of at least one reference station antenna, they did not become smaller at higher elevation angle. These biases are observable in an anechoic chamber but more difficult to characterize in operation. They may result from intrinsic antenna design as well as manufacturing variation.

While the particular orientation of each antenna and bias may be random, it is also static. Therefore, there may exist some points in the service volume where the biases tend to add together coherently consistently. Thus, these locations will experience this effect day after day. To protect these regions, the biases should be treated pessimistically as though they are all nearly worst-case and coherent. Calibration may be applied, although individual variation, the difficulty of maintaining proper orientation, and the possibility of temporal changes, hamper its practicality.

Survey Errors

Errors in the surveyed coordinates of the antenna code phase center can affect users in the same manner as antenna biases. However, survey errors tend to be much smaller in magnitude and cancel between L1 and L2.

These errors can typically be lumped in with antenna bias protection terms and mitigated in the same manner.

Receiver Errors

The receivers themselves can introduce errors through false lock or other mechanisms, including hardware failure (GPS receiver, antenna, atomic frequency standard) or software design error (tracking loop implementation).

These may be mitigated through the use of redundant and independent receivers, antennas, and clocks at the same reference station (Haines et al., 2005). However, the UDRE and GIVE must still protect against small errors that may exist up to the size of the detection threshold.

Interfrequency Bias Estimation Errors

For internal use, the correction algorithms often need to know the hardware differential delay between the L1 and L2 frequencies. These are referred to as Tau group delay (Tgd) for the bias on the satellite and IFB for the inter-frequency bias in the reference station receivers. These values are typically estimated in tandem with the ionospheric delay estimation (Wilson et al., 1999). Although these values are nominally constant, there are some conditions under which they may change their value. One method is component switching, if a new receiver or antenna is used to replace an old one, or if different components or paths are made active on a satellite. Another means is through thermal variation either at the reference station or on the satellite as it goes through its eclipse season. Finally, component aging may also induce a slow variation.

The estimation process may have difficulty in distinguishing changes in these values from changes in the ionosphere. The steady state bias value and step changes may be readily observable, but slow changes comparable to the ionosphere may be particularly difficult to distinguish. Ionospheric disturbances that do not follow the assumed model of the ionosphere may also corrupt the bias estimates. The UDREs and GIVEs must bound the uncertainty that may result from such estimation errors.

Receiver Clock Estimate Errors

Similarly, the satellite correction algorithm must estimate and remove the time offsets between the reference station receivers. These differences are nominally linear over long times for atomic frequency standards. However, component replacement or failure may invalidate that model.

Nominally, these differences are easily separated; however, reference station clock failures and/or satellite ephemeris errors may make this task more difficult. The UDRE must protect against errors that may propagate into the satellite clock and ephemeris correction due to these errors. Particular attention must be paid to correlations that may result from this type of misestimation

ALL-IN-VIEW AND SUBSET GEOMETRIES

The HMI requirement is specified in the position domain, yet WAAS broadcasts values in the range/correction domain. The users combine the corrections and confidences with their geometry to form the position solution and protection level. Exactly which corrections and satellites are used is known only to the user. Therefore, how the position error depends on the residual errors is known only to the users. WAAS cannot monitor only in the position domain and fully protect its users. Errors may vary with location, causing users to have different values, and users may be using different satellites to estimate their position. A combination of position domain and range/correction domain monitoring is most efficient.

To see this effect we can look at a specific user geometry. This example was created using Stanford's Matlab Algorithm Availability Simulation Tool (MAAST), which can be used to simulate WAAS performance (Jan et al., 2001). The user has eight satellites in view as shown in Table 1. Figure 2 shows the elevations and azimuths of the satellites along with their pseudo-random noise (PRN) identification number. Table 1 also shows the PRN, elevation, azimuth, and

TABLE 1 Satellite Elevation and Azimuth Angles, Confidence Bounds, and Projection Matrix Values Both for the All-In-View Solution and Without PRN 8

PRN	***EL***	***AZ***	σ_i	s_{3i}	s_{3i} **without PRN 8**
2	45.8°	–32.3°	2.34 m	0.595	0.451
5	11.2°	–76.8°	10.10 m	0.258	0.437
6	36.6°	48.4°	2.32 m	0.162	2.005
8	9.98°	73.0°	3.74 m	1.000	—
9	61.4°	28.5°	2.03 m	–1.928	–3.087
15	32.8°	151.0°	6.89 m	–0.015	0.174
21	42.3°	–136.0°	4.83 m	0.066	–0.003
122	40.6°	120.1°	6.19 m	–0.139	0.022

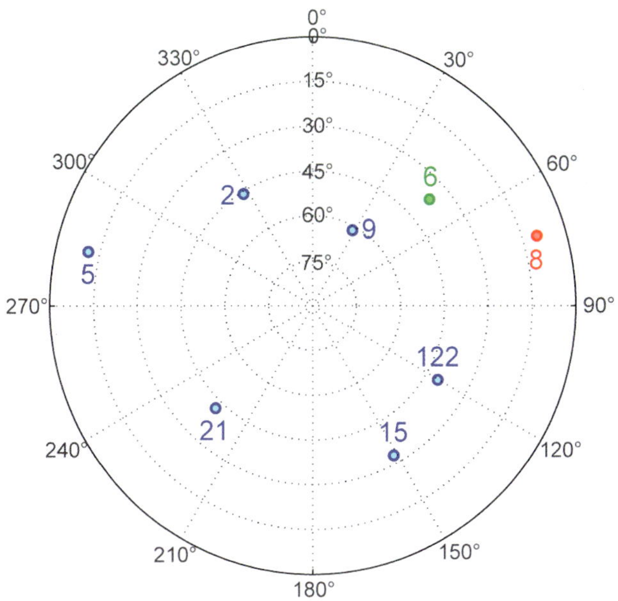

FIGURE 2 Satellite elevation and azimuth values for a standard skyplot. PRN 8 is a low elevation satellite that if not included in the solution dramatically changes the influence of PRN 6.

1-sigma confidence bound (σ_i). In addition, the fifth column shows the dependence of the vertical error to a pseudorange error on that satellite, s_{3i}. **S** is the projection matrix and is defined as $\mathbf{S} = (\mathbf{G}^T\mathbf{W}\mathbf{G})^{-1}\mathbf{G}^T\mathbf{W}$, where **G** is the geometry matrix and **W** is the weighting matrix, see Appendix J of the WAAS Minimum Operational Performance Specification (MOPS) (RTCA document DO-229D). This term multiplies the error on the pseudorange to determine the contribution to the vertical error. Thus a 1 m ranging error on PRN 2 would create a positive 0.595 m vertical error for the user with this combination of satellites and weights. The final column in Table 1 shows the projection matrix values if PRN 8, a low elevation satellite, is not included in the position solution.

With the all-in-view solution, the user has a vertical protection level (VPL) of 33.3 m (and horizontal protection level (HPL) = 20.4 m). When PRN 8 is dropped, the VPL increases to 48.6 m (HPL = 20.5 m). Both values are below the 50 m vertical alert limit (VAL) required for the localizer-precision with vertical guidance (LPV) procedure (Cabler and DeCleene, 2002). Either solution could be used for vertical guidance. Notice that the vertical error dependency changes dramatically with the loss of PRN 8. In particular, PRN 6, which had little influence over the all-in-view solution, now has a very strong impact on this subset solution. Also notice that the other values change as well. PRNs 2, 21, and 122 lose influence while PRNs 5, 6, 9, and 15 become more important. More surprisingly, the influences of PRNs 15, 21, and 122 change sign; therefore, what was a positive error for the all-in-view solution becomes a negative error for this particular subset.

The changes in the s_{3i} values with subset or superset position solutions limit the ability to verify performance exclusively in the position domain. For example, if PRN 6 had a 25 m bias on its pseudorange, it would lead to a vertical error of greater than 50 m with PRN 8 missing, but just over 4 m for the all-in-view solution. A position domain check with all satellites would not be concerned with a 4 m bias compared to a 33.3 m VPL. Thus, one would be inclined to think that no fault was present. However, the user unfortunate enough to lose PRN 8 would suffer a 50 m bias, large enough to cause harm. A 25 m bias would be more than a 10-sigma error in the range domain and thus is easily detectable. Therefore, it is the combination of range and position domain checks that protect users with different combinations of satellites. It is also possible to work exclusively in the position domain by using subset solutions; however, that approach may be numerically more intensive when considering a wide area system that must consider users throughout the service volume.

There is nothing unique about this particular geometry. To investigate how position errors can hide for one combination of satellites and be exposed for another, we set MAAST to look for subset solutions that had very different s_{3i} values in their subset solutions. We restricted the search to geometries that had VPLs below 40 m for all-in-view and then only investigated subsets with VPLs below 50 m. Of the 3,726 geometries investigated, only 2 did not change S_{3i} values by more than 40 percent.

To better illustrate the effect, the remaining 3,724 geometries had biases placed on the satellite with the largest change. Each bias was chosen such that it would lead to a 50 m positioning bias in the subset solution (a 25 m bias on PRN 6 in the example above). Each pseudorange was also assigned a zero-mean gaussian error with a standard deviation of one-half of its 1-sigma confidence bound (column four of Table 1). The broadcast WAAS confidence bounds are approximately three times larger than the nominal no-fault values (this inflation is necessary to protect against fault modes). We then calculated position errors and VPLs for both the all-in-view and subset solutions. The results are plotted in standard triangle charts (Figures 3 and 4).

Figure 3 is similar in appearance to a nominal triangle chart except the VPLs are clipped at 40 m because of our selection process and the position errors are worse than normal because of the injected error on the single satellite. However, the position errors are all below the VPL and the aggregate is not obviously biased. An observer might be inclined to believe that the system is functioning safely based on this chart. However, Figure 4 shows that the same errors and biases, but with a slightly different geometry, obviously create unsafe behavior. The subset solution removes satellites that were masking the bias for each case. The result is an obviously faulted triangle chart. Thus, a triangle chart without obvious faults, Figure 3, is no guarantee of a safe system, as evidenced by Figure 4. This has been demonstrated for real system data with much smaller biases (Sakai et al., 2010).

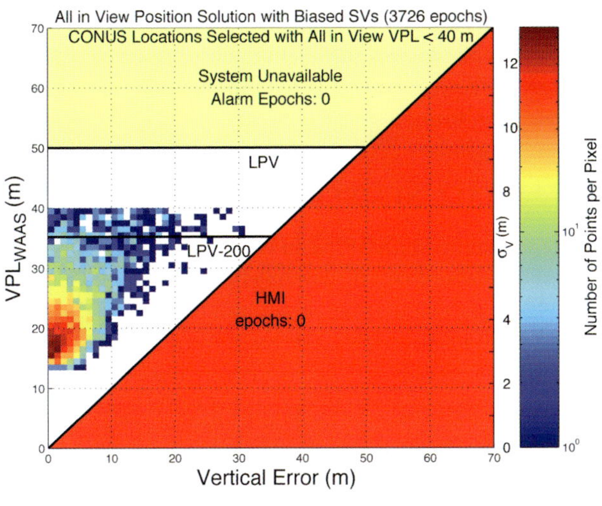

FIGURE 3 The triangle plot for all-in-view solutions including one biased satellite in each is shown. Here each bias is deweighted by the other satellites. No obvious problems are evident in this chart.

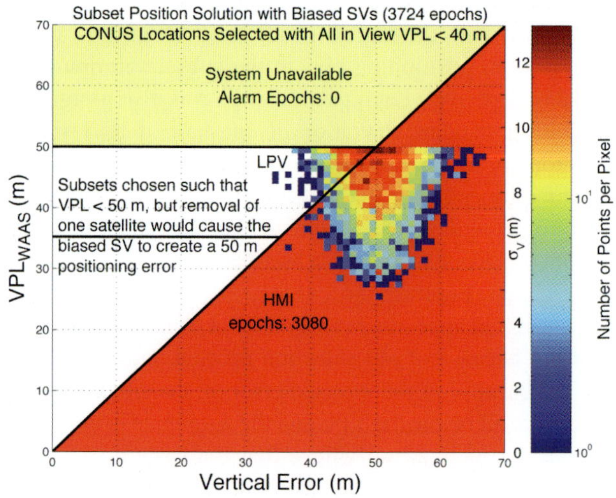

FIGURE 4 The triangle plot for the subset solutions that expose each biased satellite is shown. Here the biases are exposed as being hazardous for the user. This demonstrates the importance of checking each subset or in the range domain.

This simulation was pessimistic in its construction because the minimum unacceptable error was placed on the most sensitive satellite. On the other hand, the geometries were chosen at random and do not have any unique subset characteristics. The lesson is that it is not sufficient to observe a particular set of position solutions. The most effective method is to combine position domain monitoring with range domain monitoring.

SMALL NUMBERS AND INTUITION

The integrity requirement of 10^{-7} is an incredibly small number. In fact, it has to be; there have been more than 10^7 landings per year in the United States each year for the past 10 years.[4] Granted, only a small fraction of these are instrument landings in poor visibility; however, a larger risk value could have a noticeable effect on the overall accident rate. Furthermore, air traffic is expected to increase over the coming years. To reduce the total number of accidents while increasing the number of flights requires lowering the risk per operation. Satisfying and exceeding the WAAS integrity requirement is part of that overall strategy.

[4] Statistics gathered from National Transportation Safety Board web page, *http://www.ntsb.gov/aviation*, includes parts 121 and 135 civil commercial scheduled and unscheduled landings (does not include general aviation or military operations or unscheduled commuter operations).

It is hard to imagine the exceedingly small probability of 1 part in 10 million. By design, no individual will sample anything approaching that number of approaches. At most, an individual will sample on the order of tens of thousands of approaches. The vast majority of people will experience far fewer. Additionally, that individual will mostly experience nominal conditions and rarely the unusual events, such as ionospheric disturbances, where WAAS still has to meet 10^{-7}. Thus, personal experience is only sensitive to 10^{-4} at most. It is because so many flight operations take place under such a variety of conditions that WAAS needs to extend integrity to 10^{-7}. The full population of air travelers samples the system every year in a more thorough way than any individual can in a lifetime.

WAAS is specifically in place to protect against rare events, events that one will infrequently encounter. As a result, the situations that WAAS is designed to protect against run counter to our intuition. It is tempting to say that many of the faults listed in this paper are sufficiently unlikely to occur that we do not need to worry about them. However, when attempting to quantify the probability of occurrence we often find that it is greater than 10^{-7}. Further, because we must account for the probability of any fault occurring, specific faults are assigned sub-allocations much smaller than the full 10^{-7} allowance for all faults. Even a fault that occurs once per century has greater than a 10^{-7} chance of affecting a user in any given hour. Therefore we cannot rely solely on our observational history to convince ourselves that the system is safe.

By necessity, WAAS must work with very small numbers, probabilities of 10^{-7} and below. These probabilities are outside of personal experience and intuition. Events that seem unlikely must have an upper bound calculated for them. They should not simply be dismissed out of hand. Until one does the calculation they may not be able to distinguish between probabilities of 10^{-4} and a 10^{-7}.

CONCLUSIONS

Augmentations systems for aviation are very different from conventional differential GPS. They are supplementing and ultimately replacing proven navigational aids whose safety has been demonstrated over many years of operational experience. Consequently their safety must be proven before they are put into service. Over the course of documenting the proof of safety, the WIPP learned many important lessons. Chief among these was the use of threat models. Threat models define our fault modes, how they manifest themselves, and how likely they are to occur. They describe what we must protect against. A well-defined threat model permits a quantitative assessment of the mitigation strategy. The quantitative assessment as opposed to a qualitative assessment is essential to establishing 10^{-7} integrity.

The development and validation of the threat models is one of the most important but challenging tasks in the certification. These models are created from a combination of known physical behavior, data, and simulation. The data are very important components of this development, but it is hard to know how much data is required. The data should include all expected behaviors of the threat. The known physics can guide the initial data collection, but more data will likely be required based upon behaviors that are found in the initial set. For example, it was well-known that the ionosphere has an 11-year cycle and that it was essential to collect data from near the maximum; however, we also observed severe storm behavior that required special data collection to adequately characterize the threat.

Another key lesson is the application of the 10^{-7} integrity requirement to each approach. Rather than averaging over conditions with different risk levels, we must overbound the conditions describing the worst allowable situation. *A priori* probabilities may be used only for events that are infrequent, unpredictable, and unobservable. For example, ionospheric storms occur with certainty; therefore the system must provide at least 10^{-7} integrity while ionospheric disturbances are present. However, the onset time, exactly when the mid-latitude ionosphere will transition from a period of quiet to a disturbed state, is both rare and random. Thus, we may apply an *a priori* to that brief period of time when the ionosphere may be disturbed, but we have not yet detected it. This lesson affects how we view all of our *a priori* failure rates and probability distributions.

ACKNOWLEDGMENTS

The work for this paper was supported by the FAA Global Navigation Satellite System Program Office. The authors gratefully acknowledge the contributions from the other WIPP members.

REFERENCES

Cabler, H., and B. DeCleene. 2002. LPV: New, Improved WAAS Instrument Approach. Pp. 1013–1021 in Proceedings of ION GPS-2002, Portland, Ore., September 2002. Manassas, Va.: ION.

Collins, J.P., and R.B. Langley. 1998. The Residual Tropospheric Propagation Delay: How Bad Can It Get? Pp. 729–738 in Proceedings of ION GPS-98, Nashville, Tenn., September 1998. Manassas, Va.: ION.

Creel, T., A.J. Dorsey, P.J. Mendicki, J. Little, R.G. Mach, and B.A. Renfro. 2007. Summary of Accuracy Improvements from the GPS Legacy Accuracy Improvement Initiative (L-AII). Pp. 2481–2498 in Proceedings of the 20th International Technical Meeting of the Satellite Division of The Institute of Navigation (ION GNSS 2007), Fort Worth, Tex., September 2007. Manassas, Va.: ION.

Datta-Barua, S., T. Walter, S. Pullen, M. Luo, and P. Enge. 2002. Using WAAS Ionospheric Data to Estimate LAAS Short Baseline Gradients. Pp. 523–530 in Proceeding of ION NTM, San Diego, Calif., January 2002. Manassas, Va.: ION.

Datta-Barua, S., J. Lee, S. Pullen, M. Luo, A. Ene, D. Qiu, G. Zhang, and P. Enge. 2010. Ionospheric threat parameterization for local area GPS-based aircraft landing systems. AIAA Journal of Aircraft 47(4):1141–1151.

FAA (Federal Aviation Administration). 2009. FAA Specification Wide Area Augmentation System. FAA-E-2892C. Washington, D.C.: FAA. Available online at http://www.faa.gov/about/office_org/headquarters_offices/ato/service_units/techops/navservices/gnss/library/documents/.

Gordon, S., C. Sherrell, and B.J. Potter. 2010. WAAS Offline Monitoring. Pp. 2021–2030 in Proceedings of the 23rd International Technical Meeting of the Satellite Division of the Institute of Navigation (ION GNSS 2010), Portland, Ore., September 2010. Manassas, Va.: ION.

Gratton, L., R. Pramanik, H. Tang, and B. Pervan. 2007. Ephemeris Failure Rate Analysis and Its Impact on Category I LAAS Integrity. Pp. 386–394 Proceedings of the 20th International Technical Meeting of the Satellite Division of the Institute of Navigation (ION GNSS 2007), Fort Worth, Tex., September 2007. Manassas, Va.: ION.

Haines, B., Y. Bar-Sever. W. Bertiger, S. Byun, S. Desai, and G. Hajj. 2005. GPS Antenna Phase Center Variations: New Perspectives from the GRACE Mission. Presentation at Dynamic Planet 2005, Cairns, Australia. Available online at *http://wwwrc.obs-azur.fr/ftp/gmc/swt/Bruce/Haines_IAG.ppt*.

Hansen, A., T. Walter, D. Lawrence, and P. Enge. 1998. GPS Satellite Clock Event of SV#27 and Its Impact on Augmented Navigation Systems. Pp. 1665–1674 in Proceedings of ION GPS-98, Nashville, Tenn., September 1998. Manassas, Va.: ION.

Heng, L., G.X. Gao, T. Walter, and P. Enge. 2010. GPS Signal-in-Space Anomalies in the Last Decade: Data Mining of 400,000,000 GPS Navigation Messages. Pp. 3115–3122 in Proceedings of the 23rd International Technical Meeting of the Satellite Division of the Institute of Navigation (ION GNSS 2010), Portland, Ore., September 2010. Manassas, Va.: ION.

Heng, L., G.X. Gao, T. Walter, and P. Enge. 2011. Statistical Characterization of GPS Signal-In-Space Errors. Pp. 312–319 in Proceedings of the 2011 International Technical Meeting of the Institute of Navigation, San Diego, Calif., January 2011. Manassas, Va.: ION.

Hsu, P.H., T. Chiu, Y. Golubev, and R.E. Phelts. 2008. Test Results for the WAAS Signal Quality Monitor. Pp. 263–270 in Proceedings of IEEE/ION PLANS 2008, Monterey, Calif., May 2008. Manassas, Va.: ION.

ICAO (International Civil Aviation Organization). 2006. International Standard and Recommended Procedures, Annex 10: Aeronautical Telecommunications, Volume I: Radio Navigation Aids. Montreal, Quebec, Canada: ICAO.

Jan, S., W. Chan, T. Walter, and P. Enge. 2001. Matlab Simulation Toolset for SBAS Availability Analysis. Pp. 2366–2375 in Proceedings of ION GPS-2001, Salt Lake City, Utah, September 2001. Manassas, Va.: ION.

Jefferson, D.C., and Y.E. Bar-Sever. 2000. Accuracy and Consistency of GPS Broadcast Ephemeris Data. Pp. 391–395 in Proceeding of ION GPS-2000, Salt Lake City, Utah, September 2000. Manassas, Va.: Institute of Navigation ION).

Klobuchar, J., P.H. Doherty, M.B. El-Arinim, R. Lejeune, T. Dehel, E.R. de Paula, and F.S. Rodriges. 2002. Ionospheric Issues for a SBAS in the Equatorial Region. In Proceedings of the 10th International Ionospheric Effects Symposium, Alexandria, Va., May 2002.

Montenbruck, O., A. Hauschild, P. Steigenberger, and R.B. Langley. 2010. Three's the challenge. GPS World, July 2010. Available online at *http://www.gpsworld.com/gnss-system/gps-modernization/news/threes-challenge-10246*.

Phelts, R.E., T. Walter, and P. Enge. 2009. Characterizing Nominal Analog Signal Deformation on GNSS Signals. Pp. 1343–1350 in Proceedings of the 22nd International Technical Meeting of the Satellite Division of the Institute of Navigation (ION GNSS 2009), Savannah, Ga., September 2009. Manassas, Va.: ION.

Rajagopal, S., T. Walter, S. Datta-Barua, J. Blanch, and T. Sakai. 2004. Correlation Structure of the Equatorial Ionosphere. Pp. 542–550 in Proceedings of the 2004 National Technical Meeting of The Institute of Navigation, San Diego, Calif., January 2004. Manassas, Va.: ION.

Rivers, M.H. 2000. 2 SOPS Anomaly Resolution on an Aging Constellation. Pp. 2547–2550 in Proceedings of the 13th International Technical Meeting of the Satellite Division of the Institute of Navigation (ION GPS 2000), Salt Lake City, Utah, September 2000. Manassas, Va.: ION.

Sakai, T., K. Matsunaga, K. Hoshinoo, and T. Walter. 2010. Computing SBAS Protection Levels with Consideration of All Active Messages. Pp. 2042–2050 in Proceedings of the 23rd International Technical Meeting of the Satellite Division of the Institute of Navigation (ION GNSS 2010), Portland, Ore., September 2010. Manassas, Va.: ION.

Shallberg, K., and J. Grabowski. 2002. Considerations for Characterizing Antenna Induced Range Errors. Pp. 809–815 in Proceedings of the 15th International Technical Meeting of the Satellite Division of the Institute of Navigation (ION GPS 2002), Portland, Ore., September 2002. Manassas, Va.: ION.

Shallberg, K., and F. Sheng. 2008. WAAS Measurement Processing; Current Design and Potential Improvements. Pp. 253–262 in Proceedings of IEEE/ION PLANS 2008, Monterey, Calif., May 2008. Manassas, Va.: ION.

Shallberg, K., P. Shloss, E. Altshuler, and L. Tahmazyan. 2001. WAAS Measurement Processing, Reducing the Effects of Multipath. Pp. 2334–2340 in Proceedings of ION GPS-2001, Salt Lake City, Utah, September 2001. Manassas, Va.: ION.

Shank, C.M., and J. Lavrakas. 1993. GPS Integrity: An MCS Perspective. Pp. 465–474 in Proceedings of ION GPS-1993, Salt Lake City, Utah, September 1993. Manassas, Va.: ION.

Sparks, L., A.J. Mannucci, T. Walter, A. Hansen, J. Blanch, P. Enge, E. Altshuler, and R. Fries. 2001. The WAAS Ionospheric Threat Model. In Proceedings of the International Beacon Satellite Symposium, Boston, Mass., June 2001.

Walter, T., A. Hansen, J. Blanch, P. Enge, T. Mannucci, X. Pi, L. Sparks, B. Iijima, B. El-Arini, R. Lejeune, M. Hagen, E. Altshuler, R. Fries, and A. Chu. 2001. Robust detection of ionospheric irregularities. Navigation 48(2):89–100.

Walter, T., S. Rajagopal, S. Datta-Barua, and J. Blanch. 2004. Protecting Against Unsampled Ionospheric Threats. In Proceedings of Beacon Satellite Symposium, Trieste, Italy, October 2004.

Warren, D.L.M., and J.F. Raquet. 2002. Broadcast vs Precise GPS Ephemerides: A Historical Perspective. Pp. 733–741 in Proceedings of the 2002 National Technical Meeting of the Institute of Navigation, San Diego, Calif., January 2002. Manassas, Va.: ION.

Wilson, B., C. Yinger, W. Feess, and C. Shank. 1999. New and improved—The broadcast interfrequency biases. GPS World 10(9):56–66.

Wong, G., R.E. Phelts, T. Walter, and P. Enge. 2010. Characterization of Signal Deformations for GPS and WAAS Satellites. Pp. 3143–3151 in Proceedings of the 23rd International Technical Meeting of the Satellite Division of the Institute of Navigation (ION GNSS 2010), Portland, Ore., September 2010. Manassas, Va.: ION.

Breaking the Ice: Navigation in the Arctic

GRACE XINGXIN GAO, LIANG HENG, TODD WALTER, AND PER ENGE
Stanford University

ABSTRACT

Arctic navigation is becoming increasingly important, because oil exploration and normal shipping are both on the rise. Navigation integrity is particularly important, because an accident could be very damaging to the sensitive Arctic environment. Thus, this paper investigates the Arctic extension of space-based augmentation systems (SBAS, e.g., WAAS [Wide Area Augmentation System], EGNOS [European Geostationary Navigation Overlay Service], and MSAS [Multi-functional Satellite Augmentation System]). More specifically, it analyzes new reference stations for the far North; Iridium satellites to broadcast the integrity information to the users; and multi-constellation GNSS to improve vertical performance.

BACKGROUND

The Arctic houses an estimated 90 billion barrels of undiscovered, technically recoverable oil and 44 billion barrels of natural gas liquids according to the U.S. Geological Survey (2008). These potential oil reserves represent 13 percent of the undiscovered oil in the world. Russia, Canada, and the United States plan to explore the Arctic for extensive drilling soon. At the same time, the Arctic is becoming more accessible to normal shipping because of global climate change. New summer sea lanes have already opened up, and projections of sea ice loss suggest that the Arctic Ocean will likely be free of summer sea ice sometime between 2060 and 2080. The combination of undiscovered oil and climate change are driving a dramatic increase in the demand for navigation in the Arctic.

MORE REFERENCE STATIONS FOR SBAS COVERAGE

The Arctic is a sensitive environment, and thus navigation should have high integrity. For this reason, we are interested in extending the SBAS to serve this region. At present, none of the three operational SBAS provides meaningful service in the far North. In fact, Figure 1 shows the current SBAS availability coverage with vertical alert limit (VAL) equal to 35 meters, and horizontal alert limit (HAL) equal to 40 meters. Figure 1 is based on two of the currently operating SBAS: the U.S. WAAS and the European EGNOS. The locations of the WAAS and EGNOS reference stations are shown in Figure 2. The indicated lack of integrity coverage in the Arctic is due to too few reference stations.

Of course, we can extend integrity into the Arctic by adding reference stations to those shown in Figure 2. We include the reference stations of the Russian SDCM (System of Differential Correction and Monitoring) and Japanese MSAS systems, and add five new reference stations, whose locations are shown in Figure 3. We assume all these references stations provide the same measurement quality as current WAAS reference stations. We also assume that there is continuous user connectivity, that is, the user is always able to receive the SBAS corrections. Although SBAS GEO coverage is limited in the Arctic, there are other ways to maintain the connectivity, such as using low Earth orbit (LEO) satellites. This topic is addressed in more detail in the next section.

Figure 4 shows the horizontal and vertical availability with user connectivity and the reference stations in Figure 3. Again, we set VAL to 35 meters and HAL to 40 meters. Availability has been improved from no availability coverage to greater

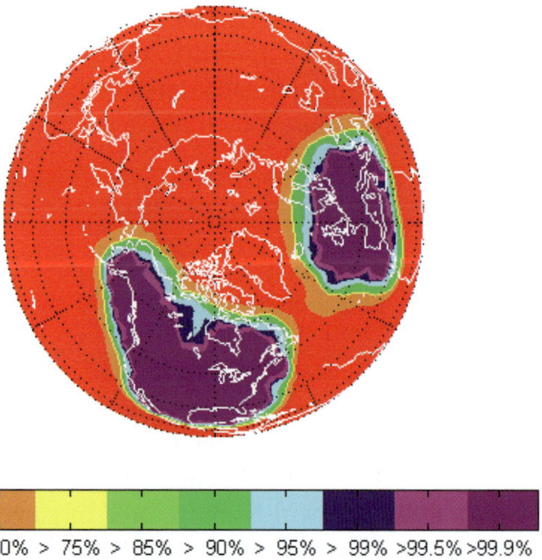

FIGURE 1 Current integrity availability in the Arctic with VAL = 35 m and HAL = 40m.

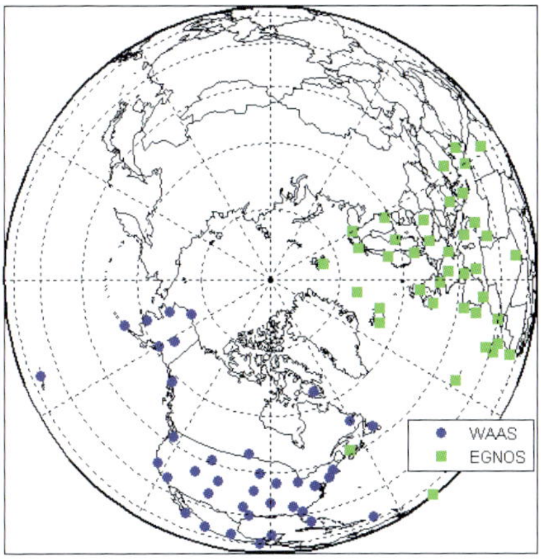

FIGURE 2 Locations of current WAAS and EGNOS reference stations.

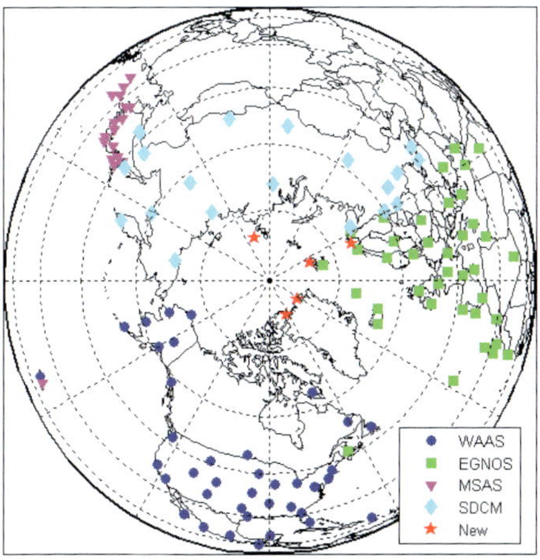

FIGURE 3 Reference stations from WAAS, EGNOS, MSAS, SDCM, and five additional stations. Source: Tyler Reid, Stanford University. Reprinted with permission.

than 90 percent coverage in the Arctic region. In the case of seafaring navigation (as opposed to airborne navigation), we can relax the vertical requirement. Figure 5 shows the availability when there is only the horizontal requirement of 40 meters. As shown, the horizontal availability exceeds 99.9 percent throughout the Arctic.

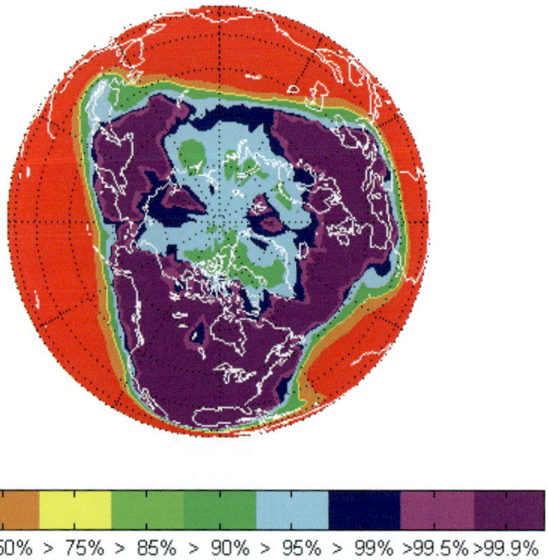

FIGURE 4 Both horizontal and vertical availability with VAL = 35 m and HAL = 40 m, assuming user connectivity and reference stations in Figure 3.

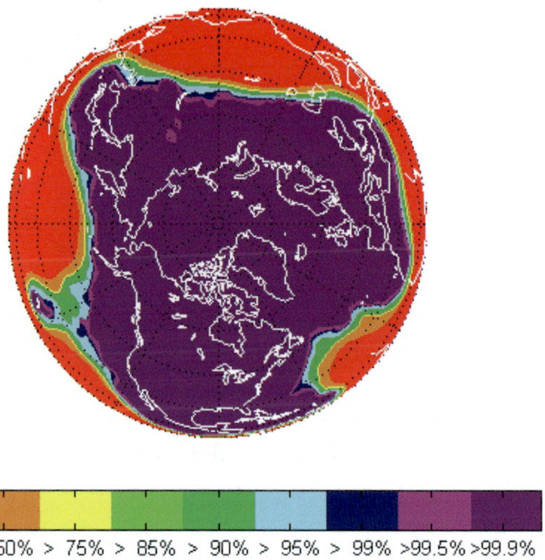

FIGURE 5 Horizontal availability with HAL = 40 m, assuming user connectivity and reference stations in Figure 3.

IRIDIUM SATELLITES FOR BROADCAST TO USERS

The second issue with ensuring integrity in the Arctic is continuous connectivity, in other words, how the SBAS messages are delivered seamlessly to the users. Currently, WAAS uses GEO satellites to broadcast error corrections to users. Because the GEO satellites orbit directly above Earth's equator, WAAS GEO coverage does not include the Arctic. As shown in Figure 6, there is lack of SBAS connectivity.

Iridium satellites are a promising alternative for communicating and broadcasting SBAS messages to the Arctic. The constellation of Iridium satellites is shown in Figure 7. This constellation includes 66 active satellites in low Earth orbit at an altitude of approximately 781 km with an 86.4° inclination (Evans, 1998). The orbiting period from pole to pole is about 100 minutes. The over-the-pole design of Iridium orbits ensures very good high-elevation satellite coverage in the Arctic. Because Iridium satellites already provide voice and data services to satellite phones and integrated transceivers all over Earth, Iridium is a strong candidate for enabling SBAS linkage to the Arctic users.

Note that the current practice of Iridium satellites is to shut off redundant beams when near the poles to avoid inter-satellite interference. Because of the low Earth orbit and thus short orbital period, the satellites are in view for only about 10 minutes. Therefore, the challenge of using Iridium satellites for communication in the polar region is to accommodate fast switching of transmitted satellites.

As a bonus, Iridium satellites could improve the vertical dilution of precision (VDOP) if the Iridium satellites also broadcast ranging signals. VDOP is a measure of how well the positions of the satellites are arranged to generate the vertical component of the positioning solution. Higher VDOP values mean less certainty in the solutions and can be caused if the satellites are at low elevations. Figure 8 shows the VDOP improvement in the Arctic using Iridium. We simulate two scenarios with and without additional Iridium satellites: using 31 GPS satellites as the healthy satellites in current GPS constellation, and 24 GPS satellites as in the original design for GPS. If we have GPS constellation only, the VDOP values in the Arctic are about 2.1 and 1.8 for 24 and 31 GPS satellites, respectively, worse than those elsewhere on Earth (1.7 and 1.5 for 24 and 31 GPS satellites, respectively) as shown in Figure 8(a,b). With added Iridium satellites, the VDOP values increase to 1.6 from 2.1 for 24 GPS satellites, and to 1.3 from 1.8 for 31 GPS satellites. Moreover, the VDOP values are more even over the Earth surface. For both 24 and 31 GPS satellites scenarios, adding Iridium satellites improves VDOPs in the Arctic. In the next section, we investigate using multi-constellation navigation to improve VDOPs.

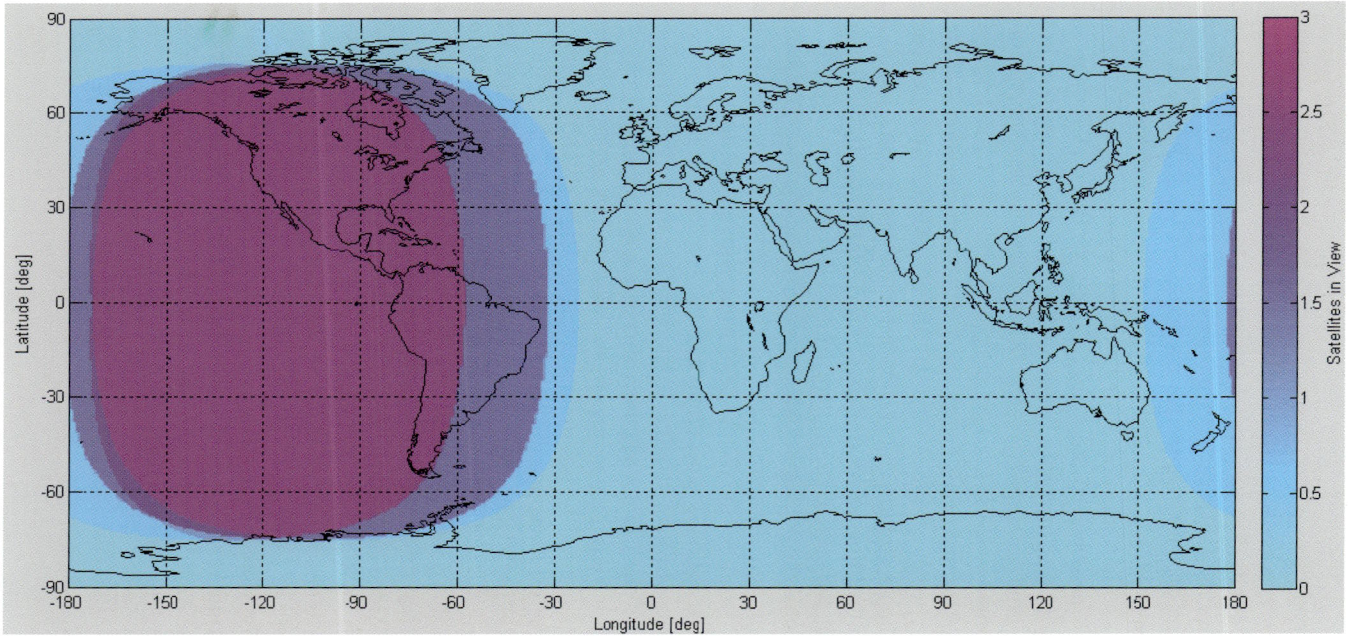

FIGURE 6 WAAS GEO satellite coverage with minimum elevation angle of 6.35°.

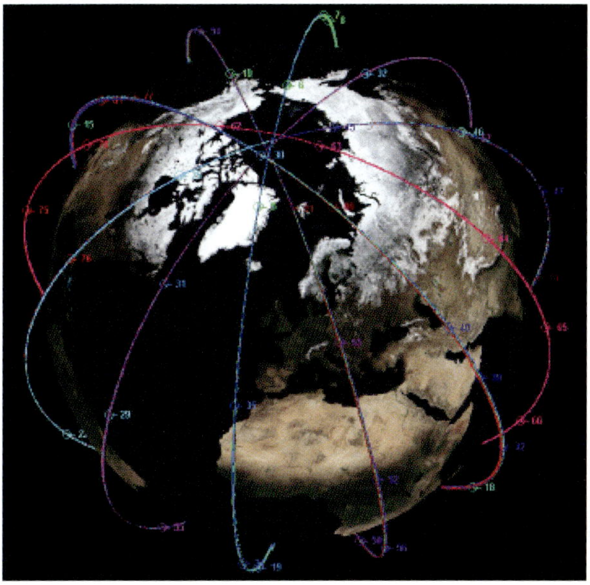

FIGURE 7 Constellation of Iridium satellites.

MULTIPLE CONSTELLATIONS FOR HIGH AVAILABILITY OF INTEGRITY

A third issue, although not as critical as the first two, is the VDOP degradation in the Arctic. Because GPS satellites are in an orbital plane of 55° inclination, there are not enough high-elevation satellites in the Arctic. For this reason, VDOPs in the Arctic are worse (i.e., higher) than those close to the equator. See Figure 9 for a comparison of VDOPs at the North Pole and at Stanford, California. In contrast, horizontal dilutions of precision (HDOPs) in the Arctic may be better than elsewhere because of this special satellite geometry.

Besides Iridium, another approach is using multi-constellation (i.e., GPS, Compass, Galileo, and GLONASS). Figure 10 shows the significant VDOP improvement in the Arctic using two or more constellations. The VDOP values reduce to below 1.3 with the help of multiple constellations. If using only two constellations, adding GLONASS to GPS is the most helpful combination. GLONASS satellites orbit at 19,100 kilometer altitude with a 64.8° inclination. Compared to the 55° inclination of the GPS orbital planes, the GLONASS constellation has better coverage in high latitudes. The VDOP improvement in the Arctic is more dramatic using three or even all four constellations.

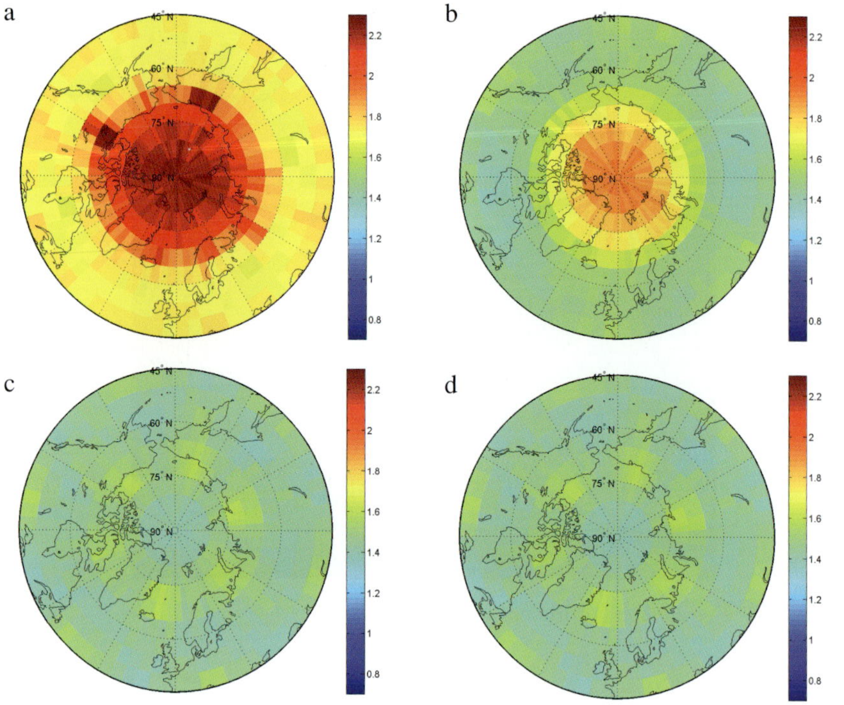

FIGURE 8 VDOP improvement in the Arctic with the help of Iridium satellites. (a) 24 GPS satellites; (b) 31 GPS satellites; (c) 24 GPS + Iridium satellites; (d) 31 GPS + Iridium satellites.

CONCLUSION

This paper identifies a need for high-integrity navigation in the Arctic and analyzes techniques to extend SBAS coverage to this critical area. We show that the current reference stations network can be augmented to provide Arctic integrity with high availability. Iridium satellites could provide a broadcast channel to the SBAS users. Multiple GNSS constellations significantly improve VDOPs and thus reduce vertical positioning errors in the Arctic.

REFERENCES

Evans, J.V. 1998. Satellite systems for personal communications. Proceedings of the IEEE 86(7): 1325–1341.

U.S. Geological Survey. 2008. 90 Billion Barrels of Oil and 1,670 Trillion Cubic Feet of Natural Gas Assessed in the Arctic. Available online at *http://www.usgs.gov/newsroom/article.asp?ID=1980*.

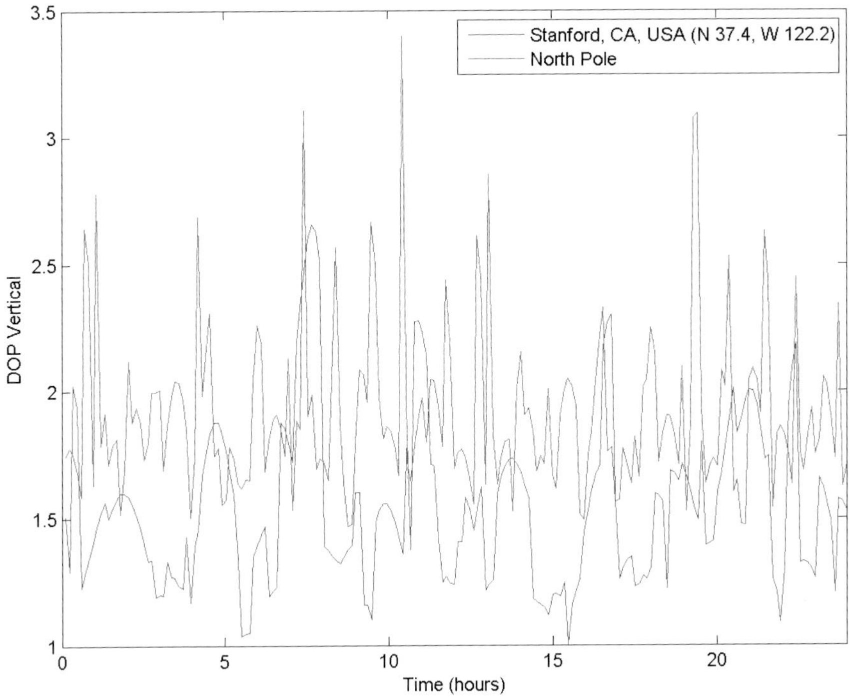

FIGURE 9 VDOP at the North Pole and Stanford, California. Source: Tyler Reid, Stanford University. Reprinted with permission.

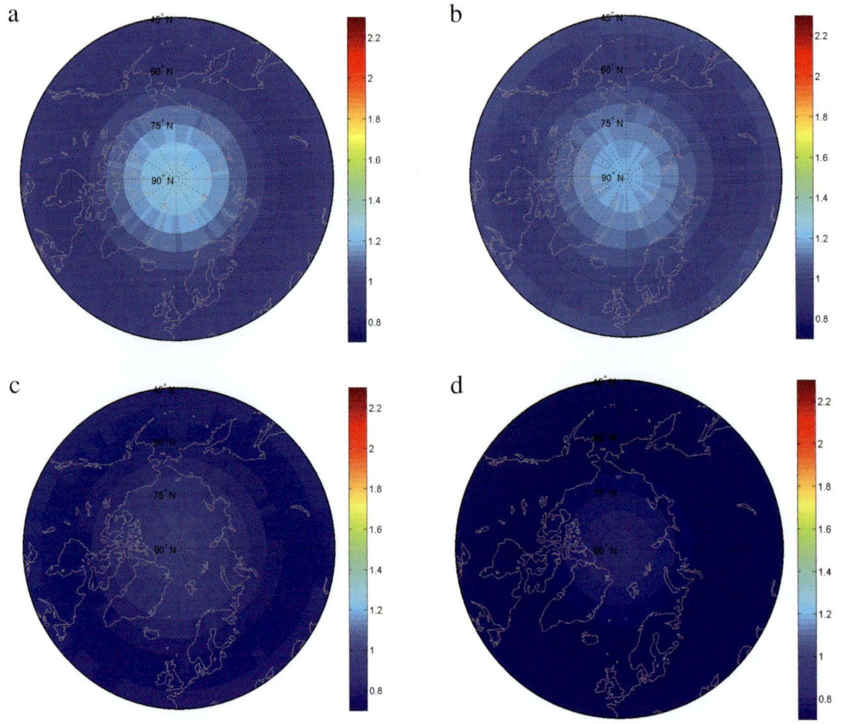

FIGURE 10 VDOP improvement in the Arctic using two or more constellations. (a) 31 GPS + 30 Compass MEO satellites; (b) 31 GPS + 24 GLONASS satellites; (c) 31 GPS + 30 Compass + 30 Galileo satellites; (d) 31 GPS + 30 Compass + 30 Galileo + 24 GLONASS satellites.

APPENDIXES

Workshop attendees.

Appendix A

Workshop Agenda

Workshop on the Global Navigation Satellite System
May 23–25, 2011
Shanghai, China

May 23 (*Monday*)

18:00–20:00	Welcome Dinner Hosted by Zhou Ji, President, Chinese Academy of Engineering

May 24 (*Tuesday*)

09:00–09:10	Opening ceremony of NAE-CAE GNSS Workshop
09:00–09:05	Opening remark by Zhou Ji, President of CAE
09:05–09:10	Opening remark by Charles Vest, President of NAE
09:10–09:20	Motivation and Overview of the Three Themes Chaired by Brad Parkinson and Liu Jingnan
09:20–09:35	Initial Comments and Observations (Free talk by any delegates)

Session I Presentations (*Chaired by Brad Parkinson and Liu Jingnan*)

09:35–10:00	Talk 1: Development of BeiDou Navigation Satellite System (by Ran Chengqi)
10:00–10:40	Talk 2: U.S. GPS Policy, Programs & International Cooperation Activities (by David Turner)
10:40–11:10	Tea & Coffee Break
11:10–11:30	Talk 3: Interoperability of BeiDou and GPS (by Lu Xiaochun)
11:30–11:50	Talk 4: Overview: The Interchangeability Problem: Signals, Coordinate Frames, and Time (by Tom Powell)

11:50–12:10	Talk 5: BeiDou Coordinate and Time System (by Yang Yuanxi)
12:10–12:30	Talk 6: A Global Safety of Life Service from Multiple GNSS Constellations (by Per Enge)
12:30–13:30	Buffet lunch

Session II Presentations (*Chaired by Brad Parkinson and Liu Jingnan*)

13:30-13:50	Talk 7: Monitoring and Assessment of GNSS Open Services (by Ding Qun)
13:50-14:10	Talk 8: Overview: The Jamming Threat (by Leo Eldredge)
14:10-14:30	Talk 9: Analyses on Development of GNSS Augmentation Technology Architecture (by Chen Jinping)
14:30-14:50	Talk 10: Impact of Intentional, Low Power, In-Band, Personal Privacy Devices (PPDs) on Aviation (by A.J. Van Dierendonck)
14:50-15:10	Talk 11: GNSS Open Signals Interference Issue and Measures (by Du Xiaodong)
15:10-15:30	Talk 12: GPS Interference Detection & Mitigation Activities (by Dave Turner)
15:30-16:00	Tea & Coffee Break
16:00-17:30	Roundtable Discussion (Chaired by Guo Shuren and Per Enge)
18:00- 19:00	Buffet Dinner
19:30-21:30	Social activity hosted by CAE for NAE delegation

May 25 (*Wednesday*)

Session III Presentations (*Chaired by Brad Parkinson and Liu Jingnan*)

09:00–09:20	Talk 13: Bei Dou Applications and Exploration (by Tan Shusen)
09:20–09:40	Talk 14: Application of GNSS to Environmental Studies (by Penina Axelrad)
09:40–10:00	Talk 15: Progress on GNSS Seismology Research (by Liu Jingnan)
10:00–10:20	Talk 16: Robotic Farm Tractors—Opportunities and Challenges (by Mike O'Connor)
10:20–10:40	Tea & Coffee Break
10:40–11:00	Talk 17: GNSS Receiver Development and Its Market in China (by Han Shaowei)
11:00–11:20	Talk 18: Attaining "Certification" of International Signals for Aircraft Landing (by Todd Walter)
11:20–11:40	Talk 19: China-U.S. GNSS Technology and Talent Exchange Channel (by Bai Shulin)
11:40–12:00	Talk 20: Arctic Navigation (by Grace Gao)
12:30–13:30	Buffet lunch

13:30–15:00	Roundtable Discussion (Chaired by Yang Yuanxi and Penina Axelrad)
15:00–15:30	Summary
15:30–16:00	Tea & Coffee Break, the end of the workshop
18:00–20:00	Buffet Dinner

Appendix B

Workshop Summary Record[1]

NAE-CAE Workshop on the
Global Navigation Satellite System
May 24–25, 2011

On May 24–25, 2011, the Chinese Academy of Engineering (hereinafter as CAE) and the U.S. National Academy of Engineering (hereinafter as NAE) held the NAE-CAE Workshop on Global Navigation Satellite Systems (GNSS) together in Shanghai, China. Mr. Wang Liheng, a member of CAE, emceed the opening ceremony of this workshop. Dr. Zhou Ji, President of CAE, and Dr. Charles M. Vest, President of NAE, attended this workshop and addressed opening remarks. The government delegates from both sides introduced the evolution process, current status, future development, applications, and international cooperation of the BeiDou Navigation Satellite system and the GPS system. In an atmosphere featuring openness, candor, and friendliness, experts and scholars conducted a frank exchange and extensive discussions on themes of GNSS compatibility and interoperability, system monitoring and service improvement, satellite navigation terminal and application technology, as well as talent exchanges.

The NAE-CAE Workshop on GNSS showed common concerns on following issues.

1. Human beings are entering a new era, with many GNSS providers offering multiple Positioning, Navigation, and Timing (PNT) services. With the development and application of multi-GNSSs, users are able to enjoy more choices and better navigation satellite services with higher accuracy, reliability, and safety. GNSS providers should enhance

[1] Joint summary prepared at the workshop under the auspices of the CAE, reflecting the observations of the workshop participants.

exchanges and cooperation, so as to improve the availability and stability of multi-GNSSs.

2. Compatibility and interoperability are beneficial to GNSS providers and receiver manufacturers and even more favorable to users. In order to strengthen compatibility and interoperability, multiple GNSS providers should explore having exchanges and cooperation, and implementing technical research together to explore all possible solutions to realize interoperability.
3. Interchangeability is a stronger and more specific goal than either compatibility or interoperability. It has already aroused multiple GNSS providers' attention because it would enable GNSS users to estimate their longitude, latitude, altitude, and time offset based on *any four satellites* chosen from the interchangeable constellations. GNSS interchangeability is more a policy issue than simply a technical issue. Unswerving efforts should be made by multiple GNSS providers on GNSS interchangeability. This is especially important in regard to safety of life applications.
4. During the GNSS construction and development process, it is essential to establish a global performance monitoring and assessing system, aiming to continuously monitor systems, improve service reliability, and improve performances. Because different GNSS providers and resources from different regions worldwide are involved in this issue, enhancing international exchanges and cooperation is quite important to realize the objective.
5. GNSS space- and ground-based augmentation can effectively improve the performance of satellite navigation and positioning services. GNSS augmentation systems have to consider accuracy, continuity, availability, and integrity simultaneously. It is mutually beneficial for the BeiDou system and GPS system to enhance exchanges and cooperation between wide area augmentation systems.
6. The BeiDou system and GPS system share the common responsibility to resist illegal interference with satellite navigation systems, which is also shouldered by governments, industry entities, and users from all countries. Interference-free environments are also the goal pursued by all users. Technical discussions can be conducted regarding interference identification, detecting, and positioning issues, while interference resistant solutions can be explored from system design, terminal applications, and other broader aspects.
7. Certification of GNSS for use in aircraft landing has already taken place for GPS L1 signals in many parts of the world. However, it remains a very challenging enterprise. As new regions seek to certify the use of GNSS, it is essential to draw on the lessons learned from previous experiences. Prototyping and careful analysis of the results are important starting points. Experience with these prototype signals and the environ-

ments they operate in is critical to develop the threat models needed to support certification. An interoperability working group exists as a forum in which space-based augmentation systems service providers can exchange ideas and experiences.

8. Satellite navigation terminals and application technologies are promoters for further development of the satellite navigation industry. Such development will be facilitated by public availability of Interface Configuration Documents for emerging GNSS systems and encouraged by private-sector engagement in receiver and application technologies. With the continuous development of GNSS application technologies, application terminals are upgrading day by day. GNSS application has gained much popularity and made significant contributions to science, engineering, and economic construction, as well as to our daily life, and is more and more indispensable for many industries. We can see a definite trend that there is still much space to improve GNSS for application technologies, and more application sectors are still waiting to be explored.

In conclusion, both sides have achieved basic identification of relevant technical issues. More mutually beneficial and fruitful exchange activities and cooperation will be conducted in the future. Both sides consider: (1) it is beneficial for technology advancement and construction of the BeiDou system and GPS system to strengthen technical exchanges and cooperation between CAE and NAE in the GNSS field; (2) it is valuable to enhance academic exchanges and mutual understanding between technical personnel on both sides to improve technologies together; and (3) such academic exchange activities shall be organized regularly or irregularly as needed.

The NAE-CAE Workshop on GNSS was held successfully in the spirit of mutual understanding and good will. Based on such an exchange platform, we hope to work together hand in hand and make further contributions to build up GNSS with higher accuracy, higher reliability, and more convenience for applications!

Appendix C

Biographical Information

UNITED STATES

PENINA AXELRAD is professor of aerospace engineering sciences (AES) with the Colorado Center for Astrodynamics Research (CCAR) at the University of Colorado, Boulder. She received her S.B. and S.M. in aeronautical and astronautical engineering, with an emphasis in avionics, from Massachusetts Institute of Technology (MIT) in 1985 and 1986, respectively, and her Ph.D. in aeronautics and astronautics from Stanford University in 1991. From 1991 to 1992 she was on the technical staff at Stanford Telecommunications in Santa Clara, California, where she worked on the development of GPS time-transfer systems, kinematic GPS algorithms, integrated GPS/inertial navigation systems, and a variety of other programs. At that time Dr. Axelrad also taught two courses at Stanford as an instructor.

In 1992 Dr. Axelrad joined the faculty in AES at Colorado as an assistant professor; she was promoted to associate professor with tenure in 1999 and to professor in August 2005. She served as associate chair of the department from 2005 to 2007 and acting chair for the 2007–2008 academic year. Dr. Axelrad has been active in research on GPS technology and primarily for space applications. She has published more than 40 technical papers and 90 conference papers and was co-editor of *GPS: Theory and Applications*. She has also been principal investigator or co-investigator on 50 research grants and contracts totaling more than $6 million. Dr. Axelrad's research interests include technology and algorithms for GPS-based orbit and attitude determination for spacecraft in low- and high-Earth orbit, multipath characterization and correction for spacecraft, aircraft, and ground reference stations, and remote sensing using GPS-based bistatic radar and occultation measurements.

Dr. Axelrad has been an active member of the Institute of Navigation (ION) since 1985, and has held several positions in the organization, including president, associate editor of *Navigation*, student session chair, space representative, Western Region vice president, chair of the ION GPS Program, and secretary of the Satellite Division. She is a fellow of ION and the American Institute of Aeronautics and Astronautics (AIAA), a senior member of IEEE, and a member of Sigma Xi. In 1994–1995 Dr. Axelrad served on the National Research Council Committee on the Future of GPS. In recognition of her contributions to the field, she has received the 1996 Lawrence Sperry Award from AIAA and the 2009 Johannes Kepler Award and 2003 Tycho Brahe Award from ION.

LANCE A. DAVIS is the executive officer of the National Academy of Engineering (NAE) . Under Congressional charter, the NAE provides advice to the federal government, when requested, on matters of science and technology. As chief operating officer of the NAE, Dr. Davis is responsible for the program, financial, and membership operations and reports directly to the NAE president.

Prior to joining the NAE, Dr. Davis served as deputy director, Defense Research and Engineering (Laboratory Management and Technology Transition) at the Pentagon from 1994 to 1999. In this capacity, he exercised oversight responsibility for the $11 billion Department of Defense (DOD) laboratory system and DOD dual-use and technology-transfer activities. He chaired the Lab Consolidation Working Group charged with restructuring the DOD laboratory system and the Affordability Task Force charged with balancing the cost/performance equation in Defense Science and Technology. Other major activities included the Quadrennial Defense Review, Lab Quality Improvement Program, Lab Diversification Program, Small Business Innovation Research, Industry IR&D, Manufacturing Science and Technology, and the Defense Technical Information Center.

Dr. Davis spent most of his career in industry at Allied-Signal Inc. (now Honeywell). He joined Allied Chemical as a research scientist in 1968 and moved through a succession of R&D management positions leading to appointment as vice president of Corporate Research and Development (R&D) in 1984. He continued in this capacity until joining DOD in 1994. As vice president of R&D, he was responsible for a corporate staff of up to 450 people with an annual expense budget in current dollars of about $100 million and a capital budget of $15 million, and engaged in research and new product development related to metals, ceramics, crystal growth, electro-optics, device fabrication, thin film deposition, polymer chemistry, engineered plastics, fibers and films, composites and biotechnology.

Dr. Davis graduated *summa cum laude* from Lafayette College in 1961 with a B.S. in metallurgical engineering. He received a master's of engineering in 1963 and a Ph.D. in engineering and applied science from Yale University in 1966. He spent two years as a postdoctoral fellow at Yale University before joining Allied. Dr. Davis is a member of Phi Beta Kappa and Tau Beta Pi. He was elected to

NAE in 1992 and received the Defense Manufacturing Excellence Award from the Multi-Association Industry Affordability Task Force in December 1999.

LEO ELDREDGE is manager of the Global Navigation Satellite Systems (GNSS) group of the Federal Aviation Administration Air Traffic Organization (ATO) responsible for the Wide Area Augmentation System, Local Area Augmentation System, GPS civil requirements, and implementation of all FAA satellite navigation programs. He has more than 15 years of experience in program management and more than 3,000 hours of flying experience with an Airline Transport Pilot rating. Mr. Eldredge has an M.S. degree in computer information systems, was recipient of the Norman P. Hayes Award from the U.S. Institute of Navigation in 2008, and is the U.S. nominated representative to the Navigation Systems Panel of the International Civil Aviation Organization.

PER ENGE is the Kleiner-Perkins Professor in the School of Engineering at Stanford University, where he is also director of the GPS Research Laboratory. The GPS Laboratory pioneers satellite-based navigation systems for aviation and maritime use, two of which are in widespread use today. The first system uses medium frequency beacons to broadcast differential GPS corrections to some 1.5 million, mostly marine, users around the globe. The second system, called a space-based augmentation system (SBAS), uses geostationary satellites to broadcast differential corrections and real-time error bounds to GPS users across continental areas. Today, SBAS supports millions of GPS users. including 70,000 aircraft during instrument approach to airports. A third system, called a ground-based augmentation system (GBAS), uses a very-high-frequency data broadcast to support aircraft landing operations. GBAS is designed to support aircraft landings in zero visibility.

Dr. Enge has received the Kepler, Thurlow, and Burka Awards and is a member of the National Academy of Engineering and a fellow of the Institute of Navigation and IEEE. In 1983, he received his Ph.D. in electrical engineering from the University of Illinois, where he designed and analyzed an orthogonal signal set for code division multiple access communications.

GRACE XINGXIN GAO is a research associate in the Department of Aeronautics and Astronautics at Stanford University. She received her B.S. in mechanical engineering in 2001 and her M.S. in electrical engineering in 2003, both from Tsinghua University, China. She received her Ph.D. in electrical engineering from Stanford University in 2008. Dr. Gao currently conducts research on arctic navigation, the Multi-constellation Global Navigation Satellite System (GNSS), and GNSS monitoring, all sponsored by the Federal Aviation Administration. She has won a number of awards, including the RTCA (Radio Technical Commission for Aeronautics) William E. Jackson Award and the Institute of Navigation Early Achievement Award, and was named one of 50 GNSS Leaders to Watch by *GPS World Magazine* in 2009.

RITA LOLLOCK is general manager of the Navigation Division at The Aerospace Corporation, which provides direct support to the U.S. Air Force GPS Program Office. Before joining Aerospace, she was employed by LTV Missiles and Electronics, where she worked on modeling and simulation, guidance and navigation problems for several inertial navigation and radar-homing programs, the Extended Range Interceptor, and the Kinetic Kill Vehicle for the Space-Based Interceptor.

Ms. Lollock joined the Engineering and Technology Division of The Aerospace Corporation in 1989. In subsequent years she transferred to the GPS Program Office as a project engineer, and she was promoted to systems director for military user equipment, chaired the Technology Assessment Panel of the Navigation Warfare Evaluation Team (NET), assumed the responsibilities of systems director for GPS modernization, was promoted to principal director for GPSIII and military applications, then principal director for system engineering in the GPS Program Office, and, in 2005, assumed her current position. She holds a B.S. and M.S. in mechanical engineering from the University of Texas at Arlington, with emphasis on control systems.

MICHAEL O'CONNOR is a recognized pioneer in the booming field of precision agriculture. As a graduate student at Stanford University in 1994, he led the team that invented the world's first farm tractor steering-control system using GPS. Upon graduation in 1997, he joined IntegriNautics (now Novariant Inc.), where he founded the AutoFarm business and delivered the world's first sub-inch, hands-free steering products to farmers in 1998. As the business grew to more than 150 employees, Dr. O'Connor held a variety of executive positions in the company, including chief technology officer, vice president of business development, and chief executive officer.

Dr. O'Connor (B.S., 1992, Massachusetts Institute of Technology; M.S., 1993, and Ph.D., 1997, Stanford University) was named to the list of Top Young Innovators by *Technology Review Magazine* in 2003 and was included in the inaugural list of 50 Faces to Watch in GPS in 2006 by *GPS World Magazine*. He was inducted into the Space Technology Hall of Fame in 2006. In 2010, Dr. O'Connor founded an independent consulting practice, O'C and Associates, to provide technology and business management services to companies working in precision agriculture, location-based solutions, and vehicle dynamics and control. Dr. O'Connor is currently a member of the board of directors for iKare Corporation and Solum Inc.

BRADFORD W. PARKINSON received a B.S. in general engineering from the U.S. Naval Academy in 1957, an M.S. in aeronautics and astronautics from the Massachusetts Institute of Technology in 1961, and a Ph.D. in aeronautics and astronautics-guidance and control from Stanford University in 1966. Currently, he is co-principal investigator and associate program manager of the NASA/Stanford Relativity Gyroscope Experiment Gravity Probe B program, a test program to validate Einstein's General Theory of Relativity using orbiting gyroscopes. He also has led a Stanford research group to develop innovative uses of the Global Positioning System (GPS) for aviation. He has been the Edward C. Wells Professor at Stanford University in the Aeronautics and Astronautics Department and the W.W. Hansen Experimental Physics Lab since 1984. He is currently a recalled Emeritus Professor.

Dr. Parkinson was vice president and general manager of Intermetrics Inc. in Cambridge, Massachusetts. He was president of PlantStar Inc., a wholly owned subsidiary that was dedicated to industrial productivity monitoring. From 1979–1980 Dr. Parkinson was vice-president for business development at Rockwell International Space Systems Group and also ran the advanced engineering and business development group. In 1978, he was a professor of mechanical engineering at Colorado State University. A distinguished graduate of the Air Command and Staff College and U.S. Naval War College, Dr. Parkinson acted as head of the Department of Astronautics and Computer Science at the U.S. Air Force Academy. In 1973, he created and then managed the Navstar GPS general program office until he retired with the rank of colonel in 1978. In 1972 he was chief engineer for the Advance Ballistic Re-Entry Program. From 1966 to 1968, he instructed astronautics, guidance, and control at the USAF Test Pilot's School at Edwards Air Force Base.

Dr. Parkinson was elected to the NAE in 1990 and has served as an NAE Councillor since July 2006. He was co-recipient of the Charles Stark Draper Prize in 2003. He has served as chair, vice chair, and member of the NAE Aerospace Engineering Peer Committee, as the peer committee chair on the Committee on Membership, and as a member on the Commission on Engineering and Technical Systems. He also has served on the National Research Council Committee for an Assessment of Precision Time and Time Interval (PTTI) Science and Technology; the Task Force on the National Research Council Goals and Operations; the Aeronautics and Space Engineering Board; the Committee on the Enhanced, Lower Cost Air Force Space Systems; and the Panel on Advanced Navigation Technology.

Dr. Parkinson is a fellow of the Royal Institute of Navigation, the IEEE (1990; Life Fellow 2008), the (American) Institute of Navigation (1999), and the American Institute of Aeronautics and Astronautics (AIAA) (1991). He also is a member of the International Academy of Astronautics and the American Society of Mechanical Engineers (ASME), and was inducted into the Silicon Valley Hall of Fame in 2007. Dr. Parkinson is the recipient of the IEEE Simon Ramo Award (2002), Kershner Award (1986), Pioneer Award (1994), and Sperry Award (1998); the Institute of Navigation's Thurlow Award (1986), Burka Award (1987), and Kepler Award (1991); NASA's Public Service Medal (1994) and Distinguished Public Service Medal (2001); Navstar Joint Program Office GPS Hall of Fame Award (1993); the Royal Institute of Navigation's Gold Medal (1983); American Philosophical Society Magellanic Premium Medal (1997); AIAA Goddard Astronautics Award (2006), Von Karman Lectureship (1996), and Aerospace

Contribution to Society Medal (2001); the ASME Gold Medal (2004); and the Defense Department Superior Performance Award for Best Program Director in the Air Force (1977). He has authored more than 50 papers on the subjects of guidance, navigation, and control and is co-editor and author of the best-selling AIAA book, *Global Positioning System: Theory and Applications*.

THOMAS D. POWELL is a systems director at The Aerospace Corporation, supporting the Engineering and Technology Branch of the GPS Directorate at Los Angeles Air Force Base. He leads the GNSS Engineering and Technology Group, which provides systems engineering support on issues of GPS constellation sustainment, satellite reliability, technology development, and spectrum management. He has supported the GPS program for more than 15 years at Aerospace, including work on handheld GPS receivers and bilateral coordination with other GNSS providers. He holds a B.S. degree in aeronautical and astronautical engineering from Purdue University, an M.S. in aerospace engineering from the University of Texas at Austin, and a Ph.D. in aerospace engineering from University of California, Los Angeles.

PROCTOR P. REID is director of the National Academy of Engineering's (NAE) Program Office. In this capacity, he oversees all NAE program activities and staff and directs the NAE policy research programs on Engineering, the Economy, and Society; Engineering and Health Care; and Engineering, Energy and the Environment. Beginning his tenure with the NAE Program Office in 1988 as an NAE fellow, Dr. Reid was appointed senior program officer in 1991, associate director in 1996, co-director in 2000, and director in 2005. Since joining the Academy, he has served as the lead professional staff to multiple NAE committee studies, workshops, and symposia on issues related to the globalization of engineering, technological dimensions of competitiveness, engineering and health care, and the future of engineering education, research, and practice. Recently he co-directed a bilateral study of opportunities for U.S.-Chinese cooperation in electricity generation from renewable resources and a consensus report on the role of noise control technology and policy in achieving a quieter environment. He is currently overseeing development of a joint National Academies-U.S. Institute of Peace roundtable on technology, science, and peacebuilding and a joint NAE-Institute of Medicine initiative on engineering and health systems.

In addition to his work with the Academy, Dr. Reid has served as secretary to the AAAS Section on Industrial Science and Technology. He has been a professorial lecturer in European studies at the Johns Hopkins University Paul Nitze School of Advanced International Studies. Before joining the NAE, he was an instructor in political economy at Oberlin College (1986–1987) and worked as a consultant to the National Research Council (1988) and the Organization for Economic Cooperation and Development (1984–1985). He received his B.A. from Dartmouth College in 1979 and his Ph.D. in international relations from Johns Hopkins University in 1989.

STUART RILEY received his Ph.D. in electrical and electronics from the University of Leeds, United Kingdom, in 1994. In 1990 he started his university research in the field of GNSS receiver development. Following his graduate work he worked for a year as a research fellow at Leeds developing a GPS/GLONASS receiver for space applications funded by the European Space Agency. In 1995 he joined Trimble Navigation Limited in Sunnyvale, California, and since then has held various engineering roles. He is currently the site director of engineering for the Sunnyvale Engineering and Construction group. He has product responsibilities for the GNSS receivers developed for the Survey, Construction, Infrastructure and High Precision OEM groups (example products are the R8 GNSS, MS992, NetR9 & BD970). He is also responsible for the core GNSS application-specific integrated circuits, and signal processing for Trimble's precision products, which are used across products including geographic information system and agriculture receivers as well as the survey, construction, infrastructure, and OEM products. He holds several patents, both issued and pending, and has published numerous technical papers on GNSS.

CHARLES R. TRIMBLE is the principal founder of Trimble Navigation Limited and served as president, chief executive officer, and chairman from 1981 to 1998. He strategically guided Trimble to its dominant role in the GPS information technology market. Mr. Trimble has been personally responsible for many of the breakthrough innovations at Trimble. For example, he holds the underlying patent on which the very successful TANS products are based. Under his leadership, Trimble grew from a startup housed above a theater to the first publicly held U.S. company engaged in providing GPS solutions. Prior to founding Trimble, Mr. Trimble had already established a reputation for innovation in development at Hewlett Packard, as manager of Integrated Circuit Research and Development at Hewlett Packard's Santa Clara division. He led important commercial advances in four areas: (1) the efficient quantization of noisy signals and their subsequent signal processing; (2) high-speed monolithic analog to digital converters; (3) ultra-high-precision single-shot digital time interval measurement techniques; and (4) establishment of the IEEE 488 bus standard. Mr. Trimble is a principal founder and current chairman of the U.S. GPS Industry Council (USGIC). He is a member of the National Academy of Engineering, the Board of Trustees of the California Institute of Technology, the Council on Foreign Relations, and the NASA Advisory Council. Mr. Trimble has served as a member of the Board of Governors for the National Center for Asia-Pacific Economic Cooperation. Mr. Trimble received his B.S. degree in engineering physics, with honors, in 1963, his M.S. degree in electrical engineering in 1964, and the Distinguished Alumni Award in 1995 from the California Institute of Technology. Mr. Trimble holds four GPS-related U.S. patents and has published articles in the field of signal processing, electronics, and GPS.

DAVID A. TURNER is deputy director, Office of Space and Advanced Technology (SAT), Bureau of Oceans, Environment, and Science, U.S. State Department. Together with the office director, he manages SAT's broad portfolio of civil and dual-use space cooperation issues and diplomatic efforts focused on multilateral science and advanced technology activities, including the development and implementation of the civil aspects of National Space Policy.

Previously, Mr. Turner was an employee of the Aerospace Corporation, serving as director of the Corporation's Center for Space Policy and Strategy, which conducts space policy and strategy analyses for government customers. He was also a senior project engineer within the Corporation, focusing on GPS policy and technology matters, such as compatibility and interoperability with other satellite navigation systems, and GPS modernization.

Mr. Turner also served in the U.S. Department of Commerce as the Executive Secretariat Director of the Interagency GPS Executive Board (IGEB), providing technical and managerial support to the IGEB and its Senior Steering Group, and forming working groups under its auspices. While in this position, he was a key member of the U.S. delegation that negotiated the June 2004 agreement between the United States and Europe on cooperation between the GPS and Galileo satellite navigation programs.

Prior to his employment with the Aerospace Corporation and the federal government, Mr. Turner was a staff officer of the Aeronautics and Space Engineering Board of the National Academies and a consultant to the Science Policy Research Division of the Congressional Research Service.

A.J. VAN DIERENDONCK (BSEE '61; MSEE '65; PhDEE '68) recently received the Iowa State University Alumni Association's Distinguished Alumni Award for being internationally recognized for preeminent contributions to their professions or life's work. Dr. Van Dierendonck made major contributions to global positioning systems (GPS) technology spanning 37 years. In particular, he is the co-inventor of the use of narrow correlator technology, which is now an industry standard for GPS receivers for multipath mitigation. He was also a major contributor in the design of the GPS L5 signal. He has received awards from the U.S. Institute of Navigation (ION) including the Burka Award (which he received twice), the Kepler Award, and the Thurlow Award. He also is an ION fellow, an IEEE fellow, and is in the U.S. Air Force's GPS Hall of Fame. He is currently the owner of AJ Systems and a partner of GPS Silicon Valley in Los Altos, California. For AJ Systems, he supports the U.S. Federal Aviation Administration, NASA, and the U.S. GPS Industry Council, primarily in GNSS Spectrum protection. For GPS Silicon Valley, he provides receivers to the international community for monitoring ionospheric scintillation.

CHARLES M. VEST is president of the National Academy of Engineering (NAE) and President Emeritus of the Massachusetts Institute of Technology (MIT).

Dr. Vest earned a B.S. in mechanical engineering from West Virginia University in 1963, and M.S.E. and Ph.D. degrees in mechanical engineering from the University of Michigan in 1964 and 1967, respectively. He joined the faculty of the University of Michigan as an assistant professor in 1968 where he taught in the areas of heat transfer, thermodynamics, and fluid mechanics, and conducted research in heat transfer and engineering applications of laser optics and holography. He and his graduate students developed techniques for making quantitative measurements of various properties and motions from holographic interferograms, especially the measurement of three-dimensional temperature and density fields using computer tomography. He became an associate professor in 1972 and a full professor in 1977.

In 1981 Dr. Vest turned much of his attention to academic administration at the University of Michigan, serving as associate dean of engineering from 1981–1986 and dean of engineering from 1986–1989, when he became provost and vice president for academic affairs. In 1990 he became president of MIT and served in that position until December 2004. He then became professor and President Emeritus.

As president of MIT, he was active in science, technology, and innovation policy; building partnerships among academia, government and industry; and championing the importance of open, global scientific communication, travel, and sharing of intellectual resources. During his tenure, MIT launched its OpenCourseWare (OCW) initiative; co-founded the Alliance for Global Sustainability; enhanced the racial, gender, and cultural diversity of its students and faculty; established major new institutes in neuroscience and genomic medicine; and redeveloped much of its campus.

He was a director of DuPont for 14 years and of IBM for 13 years; was vice chair of the U.S. Council on Competitiveness for eight years; and served on various federal committees and commissions, including the President's Committee of Advisors on Science and Technology during the Clinton and Bush administrations, the Commission on the Intelligence Capabilities of the United States Regarding Weapons of Mass Destruction, the Secretary of Education's Commission on the Future of Higher Education, the Secretary of State's Advisory Committee on Transformational Diplomacy, and the Rice-Chertoff Secure Borders and Open Doors Advisory Committee. He serves on the boards of several non-profit organizations and foundations devoted to education, science, and technology.

In July 2007 he was elected to serve as president of the U.S. National Academy of Engineering for six years. He has authored a book on holographic interferometry and two books on higher education. He has received honorary doctoral degrees from 15 universities. He was awarded the 2006 National Medal of Technology by President Bush and received the 2011 Vannevar Bush Award.

TODD WALTER received his B.S. in physics from Rensselaer Polytechnic Institute and his Ph.D. from Stanford University in 1993. He is currently a senior research engineer at Stanford University. He is a co-chair of the Federal Aviation Administration (FAA) WAAS Integrity Performance Panel focused on the implementation of WAAS. His current activities include defining future architectures to provide aircraft guidance and working with the FAA on the implementation of dual-frequency WAAS. Key early contributions include: prototype development proving the feasibility of WAAS, significant contribution to WAAS MOPS, and design of integrity algorithms for WAAS. He is a fellow of the ION and serves as its president.

CHINA

BAI SHULIN is deputy director of the Advanced Technology Institute, Peking University, and deputy director of the Joint Center of GNSS, Ministry of Education, China. He received a bachelor's degree in 1983 from the Department of Materials Engineering and a master's degree in 1986 from the Department of Engineering Mechanics, both of Dalian University of Technology, China. In 1989, he received a DEA (equivalent to an M.S.) and, in 1993, a doctoral degree (comparable to a Ph.D.) from Ecole Centrale d'Arts et Manufactures de Paris, France. Dr. Bai's major research work is on polymer composites, including fiber-reinforced and particle-filled polymer composites and polymer blends. Material systems he has investigated include: glass fiber/epoxy, carbon fiber/epoxy, wood fiber/PP, PP/PA6 blends, nano $CaCO_3$/HDPE, nano SiO_2/epoxy, steel fiber/PA, steel fiber + carbon nanotubes/PA, glass fiber/PP, etc. Dr. Bai has published more than 70 papers.

BAI YAN is associate research fellow in the Navigation and Communications Department of the National Time Service Center and leader of the "Western Light" Program of the Chinese Academy of Science. Her research has focused on navigation signal systems and signature analysis. Previously, Dr. Bai was the lead researcher in the Main Direction Program of Knowledge Innovation of the Chinese Academy of Science, a project supported by the National High Technology Research and Development Program of China. She has published more than 10 papers in leading journals.

CHEN JINPING received his Ph.D. in geodesy from Zhengzhou Surveying and Mapping College in 2001. He is currently a senior engineer at Beijing Global Information Application and Development Center. His current research includes designing integrity algorithms for GNSS and defining future integrity architectures for COMPASS.

DING QUN is a senior engineer in Xi'an Research Institute of Navigation Technology engaged in studying satellite navigation receivers, DGNSS technology, and satellite navigation application systems. He received his bachelor's of information engineering from Xidian University and his master's of communications and information from Beijing University of Aeronautics and Astronautics.

DU XIAODONG is director of the Satellite Navigation Department at Beijing Research Institute of Telemetry, where he is responsible for department management, project planning, and strategic decision making. He leads 11 GNSS engineering and technology groups (110 people), working in the design and use of GNSS receivers and payloads. Mr. Du holds an M.S. in information and communication engineering from Beijing University of Aeronautics and Astronautics and has been working in the field of satellite navigation for 17 years.

HAN SHAOWEI received his Ph.D. in satellite geodesy and navigation from the University of New South Wales and his master's and bachelor's degrees in satellite geodesy and navigation from Wuhan University. He is the founder and CEO/president of Unicore Communications Inc., Beijing, China. The company is developing GNSS chips and OEM boards that have BeiDou/Compass functionality. Previously, Dr. Han was vice president of Location Technology and SoC Engineering at SiRF Technology Inc.; vice president of Engineering and Advanced Technologies in Centrality Communications before merging with SiRF; and principal scientist and Advanced Technology Group manager at Thales Navigation (formerly Magellan). In addition to his industry experience, Dr. Han was a senior lecturer at the University of New South Wales, Australia, and associate professor at Wuhan University (formerly Wuhan Technical University of Surveying and Mapping). His research on navigation has contributed to more than 140 publications and more than 10 patents.

HAN TAO is a research assistant in the Navigation and Communication Department of the National Time Service Center, Chinese Academy of Science. Previously, he was the lead researcher on a project supported by the National High Technology Research and Development Program of China, National Natural Science Foundation of China. His main fields of investigation are navigation algorithms, data processing methods, GNSS interoperability, and mathematical modeling. Mr. Han received his M.S. from Northwest University in 2009 and was awarded the Outstanding Youth Paper Award of the First China Satellite Navigation Conference. He has published more than 10 journal papers.

LIU JINGNAN is an expert in geodesy and surveying engineering, with a special focus on GNSS technology and applications. He graduated from the former Wuhan College of Surveying and Mapping and received a master's degree in 1982. In 1999, he was elected Academician of the Chinese Academy of Engineering. From August 2003 to November 2008, he was president of Wuhan University, and since 1998, he has been director of the National Engineering Research Center for Satellite Positioning Systems. Currently, he is the 11th member of the National Committee of the Chinese People's Political Consultative Conference.

Over the past few decades, Dr. Liu has been engaged in research on geodetic theories and applications, including the establishment of a national coordinate system, GNSS technology and software development, and large project implementation. Several GNSS software systems have been developed under his leadership (e.g., comprehensive data processing software for GPS satellite-positioning systems and the establishment of provincial and city CORS systems). He has been awarded more than 10 national or provincial prizes for progress in science and technology, has published more than 150 academic papers, and has supervised more than 100 postgraduates.

LU JUN works in the Beijing Institute of Tracking and Telecommunications Technology, where her efforts have focused on the overall design and international cooperation for the BeiDou satellite navigation system. Dr. Lu's main areas of investigation are navigation frequency, signal system design, and GNSS compatibility and interoperability. She was awarded the provincial and ministerial Science and Technology Advancement Award Prize several times and has published more than 10 papers in leading journals.

LU XIAOCHUN is a professor and Ph.D. supervisor in the Navigation and Communication Research Laboratory, National Time Service Center, Chinese Academy of Sciences. Her research interests include satellite navigation and time synchronization. Dr. Lu led a navigation team in the China Academy of Science Knowledge Innovation Project, called China Area Positioning System (CAPS). She proposed a time synchronization method for CAPS transmitted signals and a method of CAPS satellite timing and designed the CAPS satellite signals. Another project in which she participated, Integrated Baseband Equipment, won Third Prize of National Defense Science and Technology. Her recent research has been focused on the seamless integration of navigation and positioning technology, including integrated navigation and positioning technology in the urban environment based on satellite positioning systems, digital TV positioning, ultra wide-band indoor positioning, and cell phone positioning. Since 2008, she has concentrated on the monitoring and evaluation of navigation satellite signals and signal compatibility and interoperation of GNSS. Dr. Lu has published more than 30 academic papers in domestic and foreign periodicals and owns several patents.

NIE JUNWEI received a B.S. and M.S. from National University of Defense Technology, Changsha, China, in 2005 and 2007, respectively. As a Ph.D. candidate, his interests are focused on signal array processing and techniques for suppressing interference in GNSS receivers.

TAN SHUSEN, deputy chief designer of the BeiDou Navigation Satellite System Project, has conducted research on navigation satellite systems for more than 20 years and hosted the research and construction of the ground application system of the BeiDou Navigation Satellite Demonstration System and China Navigation Satellite Augmentation System. He has been awarded one item of the first and one item of the second order awards for national progress and has published several monographs, including "Engineering of Satellite Navigation and Positioning" and "The Comprehensive RDSS Global Position and Report System."

WANG LIHENG, research fellow, is an expert in missile propulsion technology and aerospace engineering management. He is a past chief engineer and vice minister of the Ministry of Aerospace Industry, executive vice president of China Aerospace Corporation, and president of China Aerospace Science and Technology Corporation (CASC). Currently, he is the academician of the International Academy of Astronautics and Academy of Engineering. Dr. Wang led the development of China's first solid motor for a coastal-defense missile, a technological breakthrough. As vice commander of China's Manned Space Flight Program from 1999 to 2002, he was responsible for the development and testing of the manned spaceship and launch vehicle and the successful launch and recovery of the Shenzhou-1 and -2 spaceships. He also was responsible for the maiden flight of key defense equipment, including five new kinds of satellites. Dr. Wang's innovations in management and adaptation to changes in the economic system have contributed to the sustainable development of the aerospace industry. He has received two national awards for progress in science and technology.

WANG XUE works in the Navigation and Communication Department of the National Time Service Center, Chinese Academy of Science. His main areas of research are: navigation frequency and signal system design, signal system validation, GNSS space signal quality assessment, and GNSS interoperability. Dr. Wang received his master's degree in communication and information engineering in 2007 and his Ph.D. in celestial bodies and celestial mechanics in 2011 from Xi'an University of Electronic Science and Technology.

YANG YUANXI is professor of geodesy and navigation at both Xi'an Research Institute of Surveying and Mapping and the China National Administration of GNSS and Application. He received his S.B and S.M. in geodesy from Zhengzhou Institute of Surveying and Mapping in 1980 and 1987 and his Ph.D. in geodesy from the Institute of Geodesy and Geophysics of the Chinese Academy of Science in 1991. From 1991 to 1998, he was associate professor and then full professor at Zhengzhou Institute of Surveying and Mapping. Until 2009, he was deputy director and chief engineer at the Xi'an Research Institute of Surveying and Mapping.

Dr. Yang's international experience includes a year as a visiting scholar at the Center for Space Research, University of Texas (1995), and a year as research scientist (Humboldt Fellow) at the Institute of Theoretical Geodesy at Bonn University, Germany (1996–1997).

In 2007, Dr. Yang was honored as an "Academic Member of the Chinese Academy." He is co-author of three monographs and the author of two monographs and more than 200 papers. He has also received two items of the second order national progress awards of science and technology and one item of the third order national nature science progress award.

ZHOU JI, president of the Chinese Academy of Engineering (CAE), specialized in mechanical engineering. He graduated from Tsinghua University in 1970 and received his Ph.D. from the State University of New York at Buffalo in 1984. A professor at Huazhong University of Science and Technology (HUST), Dr. Zhou successively served as president of HUST, director-general of the Hubei Provincial Department of Science and Technology, mayor of Wuhan city, and minister of education. He was elected a member of the Chinese Academy of Engineering in 1999. During his career, Dr. Zhou was actively involved in research and development of optimal design, computer-aided design (CAD), and numeric-control (NC) technology. He advanced the algorithm of direct interpolation for NC machining and the algorithm of monotonism analysis for optimization. He and his team developed NC equipment and software packages on mechanical CAD, which have been widely used in the fields of machinery, aeronautics, astronautics, and energy. Dr. Zhou has written 11 books, published more than 200 papers, and has been honored several times with the State Award for Science and Technology Progress.

Appendix D

Acronyms

ABAS	aircraft-based augmentation system
APNT	alternate position, navigation, and time
ARAIM	advanced receiver autonomous integrity monitoring
ARNS	Aeronautical Radio Navigation System
BD NAV	BeiDou navigation
BDC	BeiDou Coordinate System
BDT	BeiDou system time
BOC	binary offset carrier
BPS	bits per second
BT	BeiDou time
BTRF	BeiDou Terrestrial Reference Frame
CAE	Chinese Academy of Engineering
CBOC	composite binary offset carrier
CDMA	code division multiple access
CGCS 2000	China Geodetic Coordinate System 2000
CL	civilian long length code
CM	civilian moderate length code
CNAGA	China National Administration of GNSS and Applications
CONUS	conterminous United States
CORS	continuously operating reference stations
CS	clock set
CTRF 2000	China Terrestrial Reference Frame 2000

DME	distance measuring equipment
DPE	data processing element
EGONOS	European Geostationary Navigation Overlay Service
FAA	U.S. Federal Aviation Administration
FSL	free space loss
GAGAN	GPS and geo-augmented navigation system
GBAS	ground-based augmentation system
GEO	geostationary Earth orbit
GGS	GPS Geodetic System
GIVE	grid ionospheric vertical error
GLONASS	GLObal NAvigation Satellite System (Russia)
GM	Gravitational Mass
GNSS	Global Navigation Satellite Systems
GPS	Global Positioning System
GPST	GPS Time
GRS	Geodetic Reference System
HAL	horizontal alert limit
HDOP	horizontal dilution of precision
HMI	hazardously misleading information
HPL	horizontal protection level
ICAO	International Civil Aviation Organization
ICD	interface control document
ICG	International Committee of GNSS
IFB	inter-frequency bias
IFR	instrument flight rules
iGMAS	international GNSS Monitoring and Assessment System
IGS	International GNSS Service
ILS	instrument landing systems
IME	inter-measurement element
IMU	inertial measurement unit
INS	inertial navigation systems
IPP	ionospheric pierce point
IRNSS	Indian Regional Navigation Satellite System
IRU	inertial reference units
ITRF	International Terrestrial Reference Frame
ITRS	International Terrestrial Reference System
ITU	International Telecommunication Union

LAAS	Local Area Augmentation System
LEO	low Earth orbit
LPV	localizer-precision with vertical guidance
LSQ	LightSquared Corporation
MCS	master control stations
MEO	medium Earth orbit
MOPS	Minimum Operational Performance Standards
MTBF	mean time between failures
MTTR	mean time to restoration
MS	monitor stations
MSAS	Multi-Functional Satellite Augmentation System
NAE	National Academy of Engineering
NANU	GPS Notice Advisory to Navstar Users
NAS	National Airspace System
NextGen	Next Generation Air Traffic Management System
NRC	National Research Council
NTSC	National Time Service Center
NUTC	national universal time coordinated
OBAD	old but active data
OCE	outer-comparison element
PANDA	position and navigation system data analyst
PDOP	position dilution of precision
PE	positioning error
PL	protection level
PNT	positioning, navigation, and timing
PPD	personal privacy device
PPP	precision point positioning
PRN	pseudo random noise
PVT	position, velocity, and time
PWV	precipitable water vapor
QZSS	Quasi-Zenith Satellite System
RAIM	receiver autonomous integrity monitoring
RDSS	Radio Determination Satellite Service
RFI	radio frequency interference
RMS	root mean square
RNAV	radio navigation
RNP	required navigation performance

RNSS	regional navigation satellite service
RO	radio occultation
RP	relative positioning
RTCA	Radio Technical Commission for Aeronautics
SARPS	standard and recommended procedures
SBAS	space-based augmentation systems
SGE	signal generation element
SLR	satellite laser range
SOL	safety of life
SoW	second of week
TDOP	time dilution of precision
TFS	Time and Frequency System
Tgd	Tau group delay
TMBOC	time multiplexed binary offset carrier
US	upload stations
UAT	universal access transponder
URAE	user range acceleration error
UDRE	user differential range error
URE	user range error
URRE	user range ratio error
UTC	Coordinated Universal Time
VAL	vertical alert limit
VLBI	very long baseline interferometry
VDOP	vertical dilution of precision
VOR	VHF omni-directional range
VPL	vertical protection level
WAAS	Wide Area Augmentation System
WAM	Wide Area Multilateration
WGS	World Geodetic System
WIPP	WAAS Integrity Performance Panel
WN	week number
WRS	WAAS reference station